CW01498984

Contents

On Thursday 9th December, leaders of the CLA, the NFU, the Tenant Farmers' Association and the National Federation of Young Farmers' Clubs met the Minister of Agriculture to confirm the agreement reached by all four organisations for reform of the law governing farm tenancies. The picture shows (l to r): Michael Halford — Chairman of the TFA: Hugh Duberly — President of the CLA; Jennifer Rabjohns — Chairman of Council for the NFYFC; The Rt.Hon. Gillian Shephard MP — Minister of Agriculture; David Naish — President of the NFU.

PENS TO PLOUGHSHARES

Michael Halford

ARTHUR H. STOCKWELL LTD.
Torrs Park   Ilfracombe   Devon
Established 1898
www.ahstockwell.co.uk

Cover illustration — MJH and Sophie.

ISBN 0 7223 3605-5
Printed in Great Britain by
Arthur H. Stockwell Ltd.
Torrs Park   Ilfracombe
Devon

Chapter 1

Nightmares

The pressure on us at the underwriting box in Lloyd's was enormous; the double-sided filing cabinet with nearly a hundred metal drawers shooting back and forth as if propelled by rocket fuel, represented a major hazard to the nose or eye. I still find myself shouting through the empty cavity, "Have you got South State Street?" to the unseen figure who had stolen it at the other side of the huge desk, the box. The answer, in the form of a metal missile, threatens to take my head off.

However, I wake up and find to my great relief that the problem is not South State Street, Chicago at all, but whether the wheat has dried out enough to consider getting the combine out before midday; no less worrying, of course, but a problem that has to be faced and dealt with on a more practical level in waking hours.

The days, in fact years, spent as an entry boy in the Sturge underwriting box, were not all the stuff of nightmares, but stressful enough. The four of us each devilled for an underwriter, and to keep us on our toes and give us wider experience, we were revolved around, and a culture of competition grew up which had us desperately trying to outshine each other. The underwriters in 'my' box, the American business, were a varied bunch; there was the brilliant Julian, who thought and wrote risks at the speed of light, frequently at lunch or in the office out of our control. He appeared to be susceptible to a good meal, show, and other blandishments, and we found ourselves presented with a stack of brokers' slips which would set us back an hour in our efforts to keep up with the entry system, and which would arouse doubt and dismay at the terms our master had quoted.

I well remember being entrusted with assessing the risk on the entire school system of British Columbia: hundreds of buildings, all of wood naturally, and representing a huge amount of business (and some hazard!). Of course they wouldn't all explode into flames at once, but some would each year, and so from the scale of rates for that class of building I halved the premium and then halved it again. The resulting recommendation, which I passed on to my boss, looked astonishingly cheap and generous to the point of folly; but no! With a wave of his magic wand, Julian slashed my figure to ribbons and I burst into tears! (Not quite, dear reader! I was made of sterner stuff then.)

Was that any better than the easier-going but plonkingly dull Cyril; kindness itself but oh so boring, and able to produce a huge volume of equally boring slips to be entered on the post cards that were the only record that we had of millions of dollars worth of insurance. He wrote the dull, routine stuff by the bucketful, and probably kept the business afloat as Julian soared into the stratosphere. In between the two we had the extremely cantankerous Jack, for whom we could do nothing right, and another eager-to-please would-be Julian, who was seeking to make his mark.

Around the outside of the hive of activity the hosts of Midian prowled around the waiting brokers, demanding their slips back before we had entered them, breathing heavily down our necks and making little sallies to try to jump the queue. By the end of the day our eyes were red, our suits were polished and our shoes scuffed. We were fit for nothing much, and no wonder the nightmares kept coming back for years after.

The other half of the business, our sister box, handled the rest of the world, and seemed to manage to live in a much more relaxed atmosphere; dare I say that they were altogether more gentlemanly in their behaviour and dealings. As a result they did not have nervous breakdowns, and we rather looked down on them; after all, we thought, the dollar business represented seventy-five per cent of the firm's business. (But we never questioned the percentage of the profit!)

The climate in the 1928 Lloyd's building had a lot to answer for; the original architects had worked on the assumption, current at the time, that hot air rises, and that if fresh air was fed in at ground level and the waste extracted from the heights, this would keep the climate sweet and fresh. They reckoned without the brokers, who complained of the draught of cold air chilling their ankles at every underwriting box as they waited. At their insistence the natural laws of physics were suspended and reversed, with cold air blown in at roof level, and stale air extracted at the original intakes. The resulting soup, which we had to breathe did nothing for our health and tempers, and was probably largely responsible for the high drop-out rate.

Of the four of us entry boys in the American box only one stuck it out to make his fortune. The rest of us may not have been temperamentally suited to the pressures anyway. Dudley, a delightfully laid-back character, had the ability to come up with a wider range of excuses for late arrival than anyone I have ever met. It was quite amazing what strange weather patterns applied to Reigate and its environs. Thick fog frequently enveloped that small area of Surrey and nowhere else; Reigate Hill was always the first to become impassable in snow; the level crossing got stuck shut several times (on the rare occasions when the trains were on time!) and a strange car was abandoned in Dudley's driveway to prevent his own exit. It was no great surprise to us when we heard that he intended to leave Lloyd's and set up a garden-contracting business (nor were we totally surprised when he severed the electric cable of his hedge cutter on his first job).

David, with a high-powered American wife, wanted something better from life, and left soon after I did; my mystery virus was probably something like ME and had the effect of making me permanently tired and disinclined for work. Medical opinions decided that fresh air and exercise was the answer, and unlike the young man who was told to "Go home and bugger yourself with the bellows!" this one took up farming. I had spent a considerable amount of effort in my early years at school trying to find an illness to prevent me going into the classroom, but allow me to play games, and the same quest later on proved equally unsuccessful.

Throughout my school life, I signally failed to manage the range of infectious diseases that were a regular feature of growing up: measles, chicken pox and mumps all caught me in the last day or two of term as I anticipated the holidays to come. Of course, the school felt it could not send back the infection into the wider world, and manically took our temperatures morning, noon and night. By managing to keep the thermometer from under my tongue I did once avoid the Disease Police and took home a severe bout of flu which laid my family low. They were not noticeably grateful for this, however, and in the future I lost a large chunk of my summer hols to chicken pox, and later on, at Winchester, mumps. This last effort was responsible for my lifelong refusal to have anything to do with the game of bridge. My housemaster, Deric Mallett and his wife were both players of national standing and, with the best of intentions, made me read Culbertson during my enforced stay. I now know what I have missed, but at the time it merely added insult to injury.

The Nightmares of this section's title are all retrospective, and usually connected with the previous slice of my life; for instance the last sentence of the previous paragraph conjures up my time at the Mons Officer Cadet School, from where many memories still come back to haunt me. I developed an exceedingly painful growing point on my heel, which prevented me wearing army boots and marching for a fortnight. This certainly relieved the stress from some of that regime, and was as near as humanly possible to the elusive illness that prevented work whilst allowing games. I still remember the first night at Mons, when the new arrivals were left in no doubt that unless we painted our beds that very evening, there would be no chance of our becoming officers and gentlemen. By 2 a.m. we had completed the task, but of course could not actually go to bed because of the wet paint. Another recurring nightmare is of driving round a triangular village green, in Yateley or Eversley I think, while the Bandit, alias Captain Vivian, taunted me about my map reading and my inability to tell him where this eternal triangle was.

Corporal-of-Horse Dodson on the barrack square at Combermere, Windsor ("Shit-head Dodson they call me. I Blanco my gaiters inside and out") shouted at me: "Take that bloody silly grin off your face! When you come back here as an officer, all ponced up, I'll tell 'em!" etcetera etcetera. In fact he was a good soldier and a very nice man. We were in the same squadron for a time, although he was never my troop Corporal-of-Horse, and at no stage did he reminisce about my days in his charge.

Up at the WOSB (War Office Selection Board) where we were assessed as suitable or otherwise to become officers, I was always the one chosen and ordered to walk the plank! A lot of the tests involved small groups of us performing unlikely tasks under the command of each in turn. I think I must have been the lightest, because whenever there was a chasm to be bridged, usually with a plank just too short for the job, the best that my mates could come up with was to stand all the rest on the first few feet of the plank, and expect Halford to walk across to the other side. Repeated attempts to achieve this always ended up the same way, with me falling into the bottomless pit. To the fact that I did not burst into tears and was prepared to repeat the experiment, I attribute my passing! It certainly was not due to my success in reassembling a bicycle bell, which I doubt ever rang again; nor do I think was it my five-minute lecture on the workings of Lloyd's, which seemed to have gone all right until a helpful friend asked a simple question, "Why is it called Lloyd's?" I went blank, and said bravely, "I don't know," while knowing full well in the cold light of day, that it had been Lloyd's coffee house.

As I marched up to the general to receive the silver stick of honour at the passing-out parade at the end of my time at Mons, the realization struck me that somehow one of my gaiters had escaped from its straps, and that I, in front of proud parents and everyone else, was the scruffiest cadet on parade instead of the smartest.

Later on, now a subaltern in the Life Guards, I had to walk through the barrack gates at 7 a.m., first parade time, in full evening dress, after my car had failed to see me home on its diet of Baxtrol/petrol (5-1); this was not only embarrassing but resulted in a lot of extra orderly officer duties.

At the head of my troop, I led a convoy of about twenty-four armoured cars, telling my driver, "Straight over this roundabout," and watching in horror as not only he, but all the rest did just that, carving deep grooves across the pristine turf on the new St Albans bypass. The driver concerned, Trooper Downes, was his own man, and one whom one would have expected to find amongst the officers' numbers; his financial resources were far greater than that of most of us junior subalterns. I assume that his ability with the disassembled bicycle bell was even worse than mine, leading to a comfortable time in the ranks. (His bulk would have ruled out walking the plank!) As his troop leader I had to witness his collection and return from camp sites in Hyde Park and Blackheath (during dock/meat or other strikes), when he was chauffeured away in the family Daimler, to the envy of the rest of us. My less-well-off but enterprising men resorted to hiring out the tents they were meant to be guarding, to ladies of the night, for the princely sum of 5/- (five old shillings) for the evening's work.

Trooper Downes had always appreciated the good things of life, and ensured a plentiful supply even out on exercise on Salisbury Plain. His crate of eggs, bedded down on the front of his scout car, formed the largest scrambled egg mess most of us had ever seen. His wedding at St Margaret's, Westminster, with the reception at the Ritz, was quite something, but by then the National Service divide was behind us, and I went as a friend. His time in the ranks did not seem to hold him back thereafter, and Michael joined the business world painlessly, and became Master of the Heythrop Hunt for a while. He gave many memorable parties, but his fondness for the better things in life made him a substantial figure, and an unhealthy one as well. Sadly he died many years before his time, but not before he had given his friends a lot of fun.

At the head of the squadron on parade for a major march past, I insisted that my lot marched in time to the music (i.e. left-2-3-4) when the rest of the regiment was wrongly marching off the beat (right-2-3-4). We were the only ones in step! I earned the approbation of the

Regimental Corporal Major, but it took some guts to stick to my guns. (It was the dress rehearsal for a big V-E anniversary parade through London, and I don't think my stance would have been so applauded on the day.) As I managed to swivel myself and all the intercom and radio cables with me in the turret of the Daimler armoured car, so that they became unplugged, nobody — least of all the driver — could hear when I urged him to stop reversing into some poor householders' new gateposts, and the despairing couple watched the demolition helplessly.

Not all my military career was composed of nightmares, and my two years of National Service was responsible for some of the best times of my life. The thrill of riding the waves of rolling plains in line abreast, either in armoured cars or Chieftain tanks, or firing the big guns at Lulworth and other ranges, was quite something; probably the navy feels the same exhilaration, but of course we were only playing soldiers. The stark reality of war, with its blood, guts and burning tanks, was a side that we chose to ignore, with the older members of the regiment altogether less enthusiastic and gung ho about it all!

After a somewhat shaky start when I joined the regiment at Windsor, and nobody appeared to notice me sitting forlornly in the officers' mess, I was befriended by one kind soul and things never looked back. Derek Bartlett took me under his wing, introduced me to his family, and couldn't have been kinder. From that contact flowed a host of lifelong friends, and we are still in touch regularly. There was some soldiering, of course, but I enjoyed that, and most of it came fairly naturally to anyone who had endured the rigours of an English public school (the winter of 1947 in particular). In fact, I don't think I was too bad a soldier, and was sorely tempted to sign on and make it my career, which is just what many of my mates did. Dad, in the background telling me that there was a real world out there, and a job waiting for me if I came NOW, was a powerful advocate.

I look back and wonder what would have happened to me if I had stayed in; twenty years on, there at Her Majesty's right hand was Simon, who had been two years behind me in my house at Winchester, and followed me into the Life Guards. He had progressed to being a senior major general, commanding the Royal Armoured Corps, and on and up even higher, finally becoming Comptroller of the Queen's Household. I got quite a shock when I had a day's shooting at Sandringham, as a guest of one of the tenant farmers, only to find Simon apparently in charge of the waiters as we had our tea in the ballroom. Another nightmare pops up here: I was warned not to fill my mouth up, as it would automatically trigger the arrival of, and questioning by, Someone Very Important. As the army of waiters rushed around with ever more enticing plates of profiteroles, meringues, and all my favourites, I could contain myself no longer and crammed my mouth quickly. Sure enough, immediately on cue, the Duke of Edinburgh appeared straight in front of me, and for the rest of the party went around with crumbs all over the front of his pullover. Blowing meringue over him must have made my crime at least lèse majesté, if not high treason. I very nearly made the same mistake again, when someone EVEN MORE IMPORTANT popped up apparently from nowhere and started talking to me. I like to think Her Majesty was giving me a second chance before saying "Off with his head!"

Desmond, another contemporary from even earlier prep school days, ended up in the Life Guards, knighted, a major general, and Governor General of Bermuda. Although not all climbed to the top of the military tree, plenty made themselves a useful and thoroughly enjoyable life, whilst saving the West from the Others.

We all had to do at least ten years on the reserve after our initial National Service, which gave us one or two worrying moments at the time of the Korean War and the Suez Crisis. However, the main effect on us was to give us an extra fortnight away from the underwriting box or whatever humdrum existence we had found ourselves in. I went to Germany, to Goodwood Races, sometimes in the air with the Army Air Corps, and even (shock/horror) on foot climbing mountains in Wales. One of my friends managed to lose three armoured vehicles as he tried to drive them along a long sandy beach which was not quite as firm as it looked. First the scout car got stuck, then the armoured car went down to pull it out and also got bogged, leaving the half-track to try to rescue them both. When that too stuck, all the men could do was climb up the cliff and watch the sea cover all three. Martin was invited to pay for the lost vehicles, which he was unable to do. (This was not a Household Cavalry unit, I hasten to add.) Another strayed over the border into East Germany and had his troop of four armoured cars impounded. They were only returned some days later without their radio sets and other equipment — more embarrassment all round!

When I mentioned the winter of 1947 in passing, it was because it was particularly long and hard. The jugs and basins of water froze in the dormitories, and we had a very hearty fresh-air fiend of a dormitory prefect whose regime made things worse. However the main memory of that winter is of the inter-house ice-hockey matches, which took the place of traditional sports and were huge fun; we all seemed to have access to skates, as did most people then, but how rarely could we have used them in the past fifteen to twenty years? When I finally got home for the Easter holidays, it was to find that the main road to the golf club and the Ridge at Woldingham, 800 feet up on the North Downs, was still deep under snowdrifts, which harboured several cars in their deep freeze after two months.

I enjoyed my time at Winchester, and still think it was a wonderful place to be and as good an education as it was possible to get anywhere. Apart from the outstanding surroundings, and the amazing range of extra-curricular activities that were available to us (and certainly in my case fully utilised), I really learnt how to absorb information and retain it. The resulting benefit has stayed with me for the rest of my life, and I have a truly ludicrous volume of useless information locked away there.

It was also rather gratifying to discover that an earlier Halford had carved his name on the entrance to the old cloisters in 1668; more of that later, though . . . .

The author (before the drink got to him)
and sister Rosemary.

Chapter 2

The Keen Student

How was it that I was to be found scrabbling frantically under the dressing table in the middle of the night shouting, "Sacks! Sacks! I must have more sacks!"? By then I had a wife and small daughter, and the strange behaviour did not seem strictly relevant to our domestic life.

I had acted on the medical advice quoted in the first chapter, and pestered all my farming contacts, who were not that many, for their opinions and help. To his everlasting credit and my deep gratitude, Christopher Miles offered to give me a start on his farm at Godstone, quite nearby. It was harvest time and my first day was spent riding shotgun on the combine whilst Christopher drove. This merely involved hooking sacks on to the grain spouts, taking them off when full, tying them, manhandling them to the chute, pulling the trip lever to allow them to fall gently to the ground in an orderly row across the field, and so on. All very well; easy to anyone who has done it before or to someone who knows nothing about it at all; panic stations to me, though. As the heap of empty sacks slipped over the rail, the grain sacks filled faster than I could tie them, and the rubbish overflowed. Meanwhile, the combine driver carried on regardless of the screams of dismay coming from his rear.

Combine driving gives one an attitude which amounts to, "I rule the world and you must all do as I say, and keep me and my machine moving whatever happens." I have noticed the tendency when I myself got behind the wheel, and the combine was three times the width and ten times the output of that early machine. Puncture on the grain trailer? Blocked elevators back at the store? Lunch time? No matter, the combine driver wears a fixed stare and carries on regardless, king of all he surveys.

Thank goodness the days of the heavy railway sacks (eighteen stone) were left behind, but for a long time there were old boys about only too eager to point out that they had run up and down the ladder to the granary all day with two of these on their back. I myself did master the handling of them, and could happily catch them on my back off the elevator and put them where I wanted. I never claimed to carry two though!

After for many years taking students to help at harvest, I realise what a brave and noble friend Christopher had been to me, and what a pain in the backside I must have been to him. My tractor driving was erratic to say the least, but paled beside my effort to ride my little BSA Bantam motorbike to work across country and arrive on time. There were cows to milk, calves to feed, etc., and there was I, struggling to put my bike chain back on in the mud. In desperation, one day Christopher said to me, "I will double your wages if you get a decent bike!"

'Wow' (I thought, as I bought myself a smart new bike), 'I can't take this through the mud,' and so found myself skating horizontally down the Eastbourne road several times. But both I and the Ariel Colt survived, and usually arrived together and on time.

They were happy days, and a great contrast to the commuter city life before; I felt well, was fascinated by the problems of crop and stock husbandry, and found the enormous drop in money barely noticeable. No train fares, no suits, shirts and city trappings meant very low living costs, and the glow of bonhomie peaked at 11 a.m. each day, as I thought of my past life. 11 a.m. was when the underwriting started in earnest and the panic set in, whereas on the farm I had already done half my day's work by then. Life was made even better as I managed to get back home for lunch, where my small daughter scooted around on her potty stealing the food from my plate. Things were not very different at breakfast as Christopher and Jean also had a small child, Richard, also scooting around stealing food! When I say stealing, it is perhaps rather too strong a description of a small bird with its mouth open permanently demanding food. It amounted to a happy family atmosphere!

During this first farming autumn my first son was born, and when I arrived at work and was asked by the foreman if it was a bull or heifer calf, I could only think of the farming implication, and not his kindly enquiry about my family addition. Christopher's foreman/ manager was Gussie Starmer, who had gone to war with his boss as chauffeur/groom/batman in the Surrey Yeomanry, and come through with honour and the rank of major. One day as we were hand pulling sugar beet, it occurred to us what exalted beet it was, that had the

14

attention of a Household Cavalry subaltern (me), a Black Watch captain (Christopher), and a Royal Artillery major. The sugar beet remained unimpressed and still came up covered in mud. To really bring us down to earth, we then had to fork it into trailers, haul it several miles to the railway station, and fork it onto railway wagons. Then, some time later it would arrive at the processing factory at Kidderminster. The whole business seemed to have a certain ring of inefficiency about it, but that was how it was for a large part of southern England.

Gussie's father had been head keeper at Marden Park before the war; this was a lovely estate nestling into the North Downs, and the Greenwell family owned some 12,000 acres of Surrey as well as a substantial piece of Suffolk. The power of the major landlords was forever there to see at Godstone, where stood the Railway Hotel, even though the railway itself had been routed through the hill to Oxted on the insistence of a previous Greenwell! When Sir Bernhard Greenwell died in the middle of the war his son, Peter, was missing in action, presumed killed, and the family were expecting to pay double death duties. Fortunately Peter had not been killed, but taken prisoner. Nevertheless, the Surrey estate had to be sold, and the returning Sir Peter set up home at Butley Abbey, near Orford.

This is all by the way, but Mr Starmer senior had been a fearsome figure, and on one memorable walk with my nanny in Marden Park Woods, there was a stand-up fight between them. "Dow" came from a long line of Suffolk farmers and gave as good as she got, while my sister and I, aged four and six, listened in wonder to the strange words that were issuing forth. Many years later I was teaching her to drive (at the age of seventy) when she had retired from her job as a school matron, and she told me a bit about her early life. She had been set for a career in banking, but "had become involved with a married man" and had to leave. The driving lessons came after she had bought a car, a smart little Morris Minor, and technically drove it perfectly well. This was all right as far as it went, but at 20-25 mph maximum she couldn't pass her test. My contribution was to take her on to the main runway of the disused airbase at Molesworth, where speed was not noticeable. Thundering up, down and around the perimeter tracks at over 70 mph, when it seemed as nothing, gave her the confidence to terrify the examiner once again, and pass.

All good things have to end sometime, and my year on the farm soon went, leaving me to take the next step — off to college; in fact the Royal Agricultural College at Cirencester — and this is where the Keen Student comes in!

On the short (one-year) crash course there were some ninety to a hundred blokes (all male then), and of those a high proportion knew exactly what sort of farm they were going back to. Those whose family farm was dairying did not want to bother with lectures on potatoes and sugar beet, and vice versa. Not so the Keen Students, mainly ex-services and older than the rest, who were also paying their own fees; because of this, and the fact that none of us knew where we were likely to end up farming, we felt we had to learn everything about everything.

We must have been the dread of the lecturers, as we sat in a row at the front asking "intelligent" questions, keeping them awake, but allowing the rest of the course to doze off. We took copious notes of every detail, and collared the lecturers after hours. How they put up with us I don't know, but they did, and I made some very good and lasting friends among them, as well as amongst my fellow students.

Geoffrey Craghill's talks on animal husbandry were always enlivened by his peculiar method of illustrating his points as he spoke by drawing on the blackboard behind his back. When he reached the conclusion he would stand aside triumphantly, to show us his great work, which was — a load of bollocks. He was a delightful and charming man, who knew his subject and put it over well, and we remained friends until he died, sadly and prematurely, of cancer.

Henry Fell, another star turn of the agricultural world, was the farms director, and to him I owed my economic survival, thanks to the sense he taught on farm and business management. He was noted for his very smart turn out, and had a 'good leg for a boot', according to those in the know! He too has remained a good friend to this day.

The hot gospel fervour of Jim Lockhart as he discussed the potato in sickness and health, its cooking quality and flavour at great length, was only matched by Dai Barling as he in his turn rhapsodised in Welsh about grass. What we didn't know about the humble vegetable was clearly not worth bothering about, and I got the impression that the entire crop-husbandry course was on the potato; possibly we did have the odd lecture or two on cereals, and I am wronging a very nice man and an excellent lecturer. This is meant to be a joke, Jim!

Dai's love of the Aberystwyth varieties was touching, and his knowledge impressive; I did not always have the same success with his techniques in the 22-inch rainfall of Hunts, as opposed to the 8 feet of West Wales. Dai went on to great things at the Arable Research Centre, and was known as a world authority on barley.

Poultry was covered by the bursar, and I came away from Cirencester with a clear and memorable recipe for a healthy flock: stand by the door of the poultry house when you let them out each morning, and clobber the last four out. (The active ones are the healthiest and vice versa.) To us, the Keen Students in the front row, there did appear to be a slight flaw to this approach: for how long would you have a flock at all?

The principal, Frank Garner, held a weekly session entitled 'National Aspects'. Apart from the dullness of the subject, Frank's delivery ensured that many went to sleep, and to add to the problems it was either held on Friday afternoon or even Saturday morning. There was considerable willingness to sign in for one's mates and the register would frequently record about five times the number of actual attenders, as everyone else looked after their social commitments. He did though give me a very helpful reference when I was applying for farms to rent.

My time at Cirencester, though stressful in many ways, was basically a very happy one. Thanks once again to Christopher and his widespread family, we moved into the kitchen wing of a house in Withington (one of four big houses in that most picturesque of Cotswold villages) which was owned by an about-to-retire Eton housemaster married to Christopher's aunt. It had a lovely garden, and was thought by my wife, Anne, to be an ideal place to train our first Labrador puppy. Samba, aged ten weeks, was a Christmas present to me, and proved to be the most wonderful, biddable and trainable Labrador I was ever to possess. Without any knowledge or books on dog training I had immense fun with her as we played games of hide-and-seek, and hunt-the-sock, etc. These were all part of my attempts to develop her nose and hunting instincts, and accounted for my being caught up a tree several times by startled visitors. This patient yellow figure sat on my jacket for hours, while her trainer found ever-less-probable hiding places before whistling her up. She gained the confidence and trust that I was not abandoning her, and I have never had such a steady dog since.

She also discovered the joys of swimming, and her own model of a racing dive virtually emptied the pond; running at full speed, she closed her eyes and performed the most outrageous belly flop, far out to sea. One day we were caught by the local landowner, Major Gunther, practising for the Olympics in one of his trout pools on the river; it was one of my more embarrassing moments, but the major was remarkably tolerant about it. Like pretty well every other Labrador known, she never lost her love for water, and liked it even better if I went in too; she tried to have an occasional rest on my back as I swam in the River Ouse near Huntingdon, and I feared that she would sink me.

B

Reggie Colquhoun, our landlord, looked very concerned for his carpets, etc. when he first met the puppy. Rightly so, but for the wrong reason! It was not Samba who made puppy pools and messes, but his own spaniel, who felt the need to mark territory out, and rather spiked Reggie's guns in so doing. At that point in my life I had only shot once or twice and didn't own a gun, but that was to change, and as Samba lived to a considerable age we had some great times together. She was also a highly fertile and prolific sexpot, and would wander off casually when the mood was upon her, to find herself a man. We had some super litters from her officially, and one or two equally good unofficial ones. Her last escapade was at the age of thirteen, when she found her way down to the rectory in Hamerton, seduced the rector's dog and received his blessing on the union as they were caught in flagrente delicto! The resulting one pup was her swan song, but Bessie was very much a chip off the old block and went to good friends of ours where she proved another long liver, and as lovable. The whole line were mobile waste disposal units, and one was even christened Hoover.

My old Tourer had seen better days and parts of it tended to fall off, so that it did on occasion present itself as a moving scrap yard. I remember once driving it into Cheltenham, its two front wings on the back seat, on my way to the welder who was going to make or mend the bull-like horns that were the mounting brackets. I attracted the attention of the police. Sadly, they saw these as a hazard to others, and I was up before the local bench in no time, and treated to a severe lecture on the behaviour of young students from the college, who thought they could get away with any crime, particularly those connected with motor vehicles. I felt hard done by, and reckoned I was being punished for the sins of others; one heir to half Scotland had recently crashed and killed someone on the Fosse Way nearby, at a speed said to be over 100 mph. However, my little old car served me well, and at any rate I arrived at the college with my notes intact more regularly than my friend, John, who when on his motorbike frequently left a trail of his nearest and dearest scribblings down the drive behind him.

From the ridiculous to the sublime: my mother acquired a brand-new, just-on-the-market Jaguar XK150, drophead. It was bright red, and she and the car together looked fantastic as she came down to seek approval for her new toy. I was allowed to drive it, but only on the condition that I kept below 30 mph! This proved difficult even in bottom gear. Now forty-five years later, I still have the car, and with its engine rebuilt it is as powerful as ever. 30 mph is still difficult!

At the end of our year we should, or perhaps more likely could, have known everything to be known about farming, but we certainly had a lot of notes. My own short course year had some notably high-profile achievers: David Naish, later to follow his Uncle Jim (Lord Netherthorpe) as President of the NFU; Bill Crossley, now Master of the Queen's Horse as Lord Somerleyton; Mike McEwen, successful Dorset farmer and MFH; Arthur Moore, the Queen's advisor on bloodstock, and many more who have pioneered and farmed their way to fame and glory. I'm not sure that all this can be attributed to the RAC short course of 1958, but it is still an impressive line up.

My own small achievement in heading the National Tenant Farmer's Association and helping to steer tenancy reform through Parliament is something I am proud of. However, I rate the fact that I farmed my land pretty well (said he modestly), leaving nearly twenty miles of hedgerow, new spinneys and a lake to show for my thirty-seven years' tenure, as my best life's work.

"Don't exceed 30 mph!"

Chapter 3

The Hunt for a Farm

So there I was, at the end of the one-year crash course, knowing all that was to be known about farming, but with very limited experience of actually doing most of it, and even less first-hand idea of the economic trials and pressures of running my own business.

More to the point, I had nowhere to go! My last months at Cirencester had been spent in a constant programme of furious deciphering and revision of my copious notes, but with at least two long hauls to farm sales or viewing days each week, and most of them leading nowhere, I was becoming slightly concerned for the future.

After attending many auctions armed with the AMC's offer to meet two thirds of their valuation of a farm, I discovered quite quickly that the base price that they were working on was so far below the starting price of the first bid, that it was an almost pointless exercise. For instance, most of the heavy land in East Anglia could be bought for £70 per acre, but the AMC valuation was in the range of £30, of which they were prepared to lend £20. The missing balance of £50 was far beyond me.

The exercise was not totally valueless to me in terms of familiarising myself with different parts of the country, and the best route to take in the overworked family Morris 1000 estate car, or the previously shamed 1938 Morris 8 Tourer. Limited to a steady 50 mph by these vehicles, I soon learnt that the faster roads were not for me, and that I could maintain my momentum and reasonable time, whilst discovering the delights of Chipping Norton, Aynho, Deddington, and Bicester on my way to East Anglia in one direction, and Wiveliscombe, Wellington and their surroundings as I scoured the West Country.

I went after many farms, all with exactly the same outcome. I have over past years revisited several of them, and followed the fortunes of the successful bidders with mixed feelings. Hobbles Green Farm in

Suffolk seems to have bankrupted at least three owners, and I am very relieved that the AMC valuer from Bidwells put an absolute veto on that one. I still regretted missing some, where they looked well farmed and profitable in spite of not having me in charge, but I can honestly say that none of those where my attempts failed were as good as the one where I ended up. In the auction room there was always a lot of speculation: "He doesn't look good for much of a bid," nudge-nudge, wink-wink, and I never managed to look nonchalantly casual and disinterested under this scrutiny.

I did acquire a fairly comprehensive knowledge of Norfolk and Suffolk in this way, while getting nowhere in terms of my primary task and, still carrying my trusty but ever more dog-eared lecture notes along too, managed to turn in some reasonable exam results. In fact I was told I had got top marks, but still had nowhere to go, and now had a wife and two and a half children to worry about.

My bids for tenancies were slightly less discouraging, in that I was short-listed for several; after hearing nothing from one of my more hopeful attempts, I wrote to Bidwells to enquire what had happened to my bid for Barnardiston Hall Farm, on the Thurlow Estate. I was told that I had not been successful, but that a new tenant for a farm on the Hamerton Estate had fallen through at the eleventh hour, and was I interested? I found it on the map, the only other possibility being in Yorkshire. At Thurlow, incidentally, 'Dow' had grown up, and her family, the Dowsetts, had been farmers for generations. She was very scathing about the 'new' owners of the estate, the Vesteys having made a lot of money in trade!

My good luck was owing to four factors; the first being that the previous young tenant had married a local landowner's daughter and was going into partnership with his father-in-law. That would not have let me into the frame had not the chosen replacement taken exception to the repairs clause in the proposed new lease for Grange Farm, and baulked at the idea of being responsible for the water main. He saw this as outrageous, and likely to be very expensive over time, but I was desperate, and in the event that clause cost me a total of under £1,000 in the next twenty years. The other two points in my favour were the good reference from Frank Garner, my Principal at the RAC, and the fact that I seemed to hit it off with Francis Pemberton when he interviewed me.

To cut a longer story short, but in time only about six weeks, I was offered and accepted the 370-acre farm without my wife even having a chance to see her future home, and was moving in on October 10th 1959, having sold our previous cosy little bungalow in the southern

21

suburbs of London in the same rush. She had to cope with the idea of a large, seven-bedroomed, early Victorian house with one bathroom, one loo (upstairs) and an outside privy. We had no central heating then of course, but who did?

I was now a farmer! HELP! The biggest boost to my confidence was when my father, who had been sceptical about the whole venture, and very disappointed when I left the City and the firm which he had risen to head, gave me a brand-new Landrover as a touching gesture of parental love and faith in me.

The old cart hovel.

Chapter 4

The Farm

I was now the proud tenant of 370 acres of Huntingdonshire which I had, in my wisdom at the first interview, proclaimed to be ideal wheat and sheep land; this had rung bells with Francis Pemberton, and it was to some degree responsible for my being offered Grange Farm.

The Hamerton Estate was the same shape and size as it had always been; of course I knew nothing of its past history at the time, but as I have become interested, and done a little bit of research, some fascinating facts have come out. The great midland forest of Bruneswald had stretched from Bedford up to Northampton and across to Peterborough, where it met the Fens. It was almost impenetrable to travellers once the Romans had left, but the evidence of the 400 years of their rule is still there in the road system and place names. Coldharbour, Caldecote and Folly are all derivations of the Roman soldiers' night leaguers, and Ermine Street runs straight up through Huntingdon on its way from London to Lincoln. The settlements and farmsteads that date from then are mainly on the easier land alongside the rivers, or rather just above the flood level! There have been heaps of Romano-British pottery found on my farm, dated to the middle of the Roman occupation, i.e. about AD 200; we found what appeared to be a rubbish tip when we were planting a spinney, and later on when a gas main was driven through the farm, aerial photography and MRI scanning located other sites, but they were not major, and did not seem to be part of a big estate.

So we assume it was the Saxons, with their pigs and axes, who really established the boundaries and cleared the bulk of the woodland; working backwards from the record of the Domesday Book in 1080, the estate of *Hambertune* was granted to Eubo of Normandy (Dapifer, the sewer to his friends — heaven knows what his bodily hygiene must have been like! Hundreds of years later Queen Elizabeth was said to have taken a bath each month 'whether she needed one or no'). The land previously had belonged to Ulfek (or Ulfkell) the Dane. The Danish invasions were in the mid-800s, and it was probably taken from Eggbald, the Saxon. The boundary hedges are as ancient as they come, with every conceivable type of hedgerow plant to provide back-up evidence. (Each species represents a hundred years, as a rough dating guide.)

The enclosure of the old-style big fields and strip farming happened very early at Hamerton, and the map of 1638, which is held in the record office in Huntingdon, shows an instantly recognizable field pattern. Two hundred years later when the whole estate was tile drained and mapped again the fields were still the same, and not much has changed today on my old farm. When I took over in 1959 the average field size was thirty acres, and that is not far from the figure of the first enclosure over three hundred years back. Other insights into the landscape changes come from a record of the planting of the Grove Wood in 1245, as a deer sanctuary, and some Victorian fox coverts which overlie the existing ridge and furrow pattern. Under the Grove Wood there is a more complex pattern of ditches, which makes it look as if it covered the site of an earlier settlement.

The land was mainly heavy boulder clay, needing three horses to pull one furrow on most of it. There were many fields which had hardly ever been ploughed, and those which had been worked deeper than four to five inches were very few. A quarter had to provide food for the power units (horses) and so when I was let additional blocks to tack on to my original 370 acres, I had to re-christen the old Horse Meadows, Barn Fields and Home Closes, which were on each holding. This resulted in some strange names, and a degree of confusion, although less so than sending tractors out to a choice of four identically named fields.

Having, in my know-all confidence, pronounced the land as ideal wheat and sheep country, it came as quite a shock to me, as well as others, that I should immediately set about growing quite a substantial acreage of potatoes; lost in the mists of time now, the reasons for this decision escape me, but I did nevertheless have some sixty acres of the crop in the ground for my first season. Jim Lockhart had a lot to answer for! So, for that matter, did Dai Barling, since so much of my grassland information had been relevant to the high rainfall of Aberystwyth, or at least the Cotswolds.

I certainly did put a lot of wheat in, but fired with enthusiasm for direct seeding of grass in the autumn, I presented the most disgracefully scruffy mixture of infant grass seedlings, heavily overwhelmed by the natural vegetation. This was the wonder of my new neighbours, who could scarcely contain their glee at the mess, and my embarrassment at it. The plus side was to follow on its heels as the shooting season arrived, and my utterly charming and delightful Irish landlord discovered a fantastic new pheasant drive where none had been before. Forty acres of rubbish attracted birds in dozens from the next estate over the boundary, and they spread out across the long half-mile expanse of pheasant heaven, flying for home with a dedication and enthusiasm which can rarely have been bettered. I had to take a grass topper to this in the end, but it was good while it lasted, and gave me a head start with Major Billy Bell, who thereafter treated me like a favoured son, and always asked me to shoot when he made his biannual visit from Fota Island, County Cork.

I came away from Cirencester with the fixed idea that theirs, if not the only way to farm, was certainly the best. My sheep had to be Cluns (not just because they had such pretty faces!) and a more idiotic and mentally unstable collection it would have been hard to find. It was several years before I finally succumbed to commonsense and local practice and got rid of them in favour of Mules and later Mashams.

The autumn of 1959 followed a really hot and droughty summer, and I found myself trying to turn concrete blocks into seedbeds for the winter wheat, which was meant to be my banker. My friend (Ian McLeod) and I spent days trundling up and down the arid clots, leaving so little impression that Ian solved his problem by tying bits of handkerchief in the hedge at either end to avoid constant repetition. This was before I had officially taken over the farm and its men, but was the right of pre-entry for a new tenant. As must have been obvious to any onlooker, the cultivations in those conditions were a complete waste of time, rubber and diesel, and the only benefit was to me psychologically, as I felt I must do something!

I did make some awful mistakes during those first years, but as time went on I came to accept the fact that I would still be making them, or different ones every year; nor would I be alone in that. Each season had a sting in the tail, and cultivations that worked for one year caused disaster or chaos in another. I collected farm machinery for every conceivable weather condition as it arrived, and the result was an enormous selection of little-used machines in my final farm sale. This is jumping ahead some thirty-seven years, but I did say this would not be a blow-by-blow account!

For my first harvest I had determined to have a drier up and running, and equally sure that it would have nothing to do with sacks (see the beginning of Chapter 2). The rather pathetic set-up that resulted had a very limited output, based on a 30-cwt/hr Alvan-Blanch; however, it was an exceptionally wet and difficult harvest. The five bins, which were all I could afford, were totally inadequate (two being self-emptying pre-drying containers anyway); what was worse was the lack of any fixed access to them, let alone a catwalk. Health and Safety men were not so prevalent then, but we all survived, and I became blasé as I walked around the tops of the bins. I daren't even think about it now!

The worst aspect of the whole set-up was the nightmare of the grain elevators; the cups, bolted on to chain and flight, constantly worked loose and blocked the system however slowly we fed it. Two harvests on and I had to change the drier for a bigger one, and threw out all the elevators at the same time; even after that there was the occasional foray into the dungeons of the elevator pits, but accidents will happen, and human error always crept in somewhere (see further on, under 'Students')!

In spite of all its initial faults, in the land of the blind the one-eyed man is king, and I found myself suddenly very popular, and drying for my neighbours; even to the extent of bagging off and working through the night to do it all. I hadn't forgotten my good resolution about sacks, but the need for cash was great at a time when the harvest price for barley was £16 per ton. It helped, but I had other worries: my combine had a very narrow and inefficient pick-up, though a wide drum, and there was one twelve-acre field of barley that I never harvested. Add to that the thirty acres of potatoes that I couldn't lift because of the wet, and you had a worried farmer. Luck played a big part in my survival that winter, as it was very mild and I was able to lift and sell seventeen tons of Majestics per acre in April. Ouch, though!

Even with the harvest conditions far from ideal, I and my young children had fun on the grain trailers; only three-tonners, and therefore quite low enough for me to throw them up over the side into the corn, and travel to and fro with them safely and happily sitting in the heap. The tractors that I used to haul the grain trailers were little grey 25-hp Fergies, which really struggled up the hills, and voiced their protest by snapping drive shafts. That couldn't happen later on with my second family (my first marriage having fallen apart); the trailers were much higher ten-tonners by then, and I was older and tireder.

Over the whole period of nearly forty years that I was farming, I watched the cycle of minimal tillage, direct drilling and continuous wheat come around with predictable regularity. Just as regular was the rediscovery of the plough, together with old-fashioned good husbandry. Each cycle sees a doubling of the horsepower available.

One of the lessons taken away from Cirencester was that five hundredweight of wheat pays for an awful lot of cultivations, and skimping on the latter was seldom worthwhile. In fact I worked towards maximising my first wheats as it became more and more apparent that there was a dramatic reduction from those down to nearly a fifty per cent drop for the third crop. This was highlighted in the record harvest of 1984, when we had a five-ton per acre crop alongside a fifty-hundredweight one of the same variety, drilled on the same day and with exactly the same treatment.

One practice that has gone for good, thank heavens, is straw and stubble burning, which aroused so much ire and broke many friendships. This was particularly so when the neighbours were not farmers, and I'm sorry to say that I did my share of polluting the atmosphere and even went into print to defend the practice. In principle it was indefensible, but at the time there was very little choice. There was a vast surplus of straw, over and above the twenty-five per cent or so needed nationally for livestock, and nobody had thought up a use for the rest. (The exception was a small quantity turned into mushroom compost.) The machinery to chop and spread had yet to be invented, and so the humble match it had to be!

The price for this was appalling damage and slaughter to young wildlife, and the cloud of smuts which covered everything throughout harvest, from householders' fresh paint to hospital air-conditioning systems. Farmers were public enemy numbers 1, 2 and 3, and earned the title of Anti-Social Buggers!

Within a couple of years a tractor-drawn chopper was on the market, and soon after followed the integral version, built into the guts of the combine itself. What joy — my life at harvest had been all firebreaks and nervous breakdowns as I waited for the wind to set fire to a standing crop; now it was the duller routine of overflowing bins, muddled varieties and lorries turning up out of the blue.

Only once did I have a field cleared for me by an act of God, when a whirlwind arrived just as the combine finished the last cut, and methodically hoovered up every single straw. As the cloud of debris ascended high into the air and drifted eastwards, I wondered who the lucky recipient would be, but it travelled so far that I never did find out. I was left with a clean field to start working, so no complaints here.

Gone long ago are the two-hundredweight railway sacks, and then we lost the one hundredweight seed corn bags and found ourselves struggling with half-hundredweight packets; even that physical pleasure is denied us, with the advent of the big one-ton bags and telescopic loaders. What wimps we have become, but maybe I can blame the need for a new hip on my having carried everything on my right shoulder for years.

First wheats meant a continous battle to find break crops that would not ruin me, hence the colourful displays of linseed, rape, beans, lupins, peas and other items which the reader may well find under the heading of Great Disasters.

Slaves, Students and Others:

I was remarkably lucky and well served by the regular staff that I had throughout my time. When I took over, my predecessor had milked cows, employing a husband and wife team to do so, but the whole set-up was antiquated and I had no inclination that way, so that was one headache and two wages less. Another man left after a week, convinced that I was bad news, when I insisted on changing the pattern of the working day which he had brought from the Fens. (The day was divided into two — 7-11 a.m. and 11.30-3.30. I did not feel that my team could perform without proper breaks or time for breakfast.) I was down to a hard core of four, of which three came from the same household. Eric Possewanne was an ex-POW, who had stayed on where he had been billeted, and married; he had worked as a cowman in East Prussia before the war, had fought and lost toes through frostbite on the Russian front, and finally been captured in Normandy. Grange Farm was a haven after that lot, but he was used to hard work and was a great asset, skilled in both stock and tractor work. His mastery of the English language was almost as bad as that of the average Pole, which is saying something. He had trouble with his back, and would spend some time each day hanging from a beam in the barn, suspended from a bicycle handlebar under his armpits.

His step-family included two working sons, and the group stayed with me until the elder of the two brothers came into some land and a farmhouse and they all moved on. Also part of this household was a teenage daughter, Marjorie, who was just right for pram-pushing, baby-sitting and that sort of thing. Forty years on she is still with me, now officially cleaning lady, but in reality a loyal and much loved family friend.

In the time of change after Eric and his family left, I had one or two who were not quite all that they seemed. One was pursued by the police for various small-scale frauds, and he turned out to be incapable of telling the truth in anything; no matter how trivial, he needed to make up something better and more exciting. (We were asked to believe that our local contractor had been chasing a girl around in a very public wheat field, both stark naked.) It was all too good to be true, and life became much duller when he flitted. Yet another was tracked down and removed forcibly by his brothers-in-law, on behalf of their sister who was his real wife. A third was up to one or two dodgy dealings, and I found myself being quizzed by the police, because my vehicle had been parked illegally in London the previous weekend. The licence number was that of my Ransome combine harvester, and it seemed unlikely to have been in Park Lane at midnight, but you never know! Another good man left when he had filled the cellar with enough empty gin bottles, and so forth.

When my original landlord decided to pass on the estate to his grandchildren, their immediate reaction was, "Sell it and give us the money!" What a waste of their inheritance, but the whole estate, and a lot more like it, was bought by the Pension Fund Property Unit Trust, and I was offered an extra 500 acres by the new owners.

Peter Hall turned up wondering if I would need any more men. Indeed I would! I also heard that Sylvia Rowley at Alconbury was concerned at the number of men she was carrying (three brothers from the Slack family), and didn't want to push anyone out. So it came about that for the last twenty-five years I was fortunate enough to employ Charlie and Peter, who worked for me as if it were their own business, and made my life so much easier and trouble free — on the farm at any rate. Both could and did turn their hand to anything, and were just as adept at lambing ewes as drilling the seed. However, as the machinery became ever more specialised and massive, each tended to take responsibility for different areas, and it became harder to swap jobs around. I, of course, became less and less able to perform usefully on machines that were so far in advance of the basics.

I always felt the need for more manpower at harvest and also, as the sheep flock expanded, for lambing as well. The result was that we accumulated a vast army of friends, who had all started life with the Halfords as slaves: some as cousins, many as the offspring of acquaintances and friends. We are still in touch with several Kiwis and Aussies, and have even been to one's wedding at the bottom end of New Zealand. Most of them enriched our lives enormously, but one constant thread ran through the lot: each in his or her turn would make some dreadful mistake, and these ranged from the comparatively minor sins of tipping loaded grain trailers into ditches, or the wrong bin of a different crop at the lower end of the scale. Up the level a bit and we had runaway trailers crashing into my car, elevated trailers ripping out the barn roof, and ditto with the fork-lift; unloading augers on the combine trying to fell the electric poles (twice by the same chap), and the most destructive when our brand-new combine (an International Axial Flow) was driven into by the equally-brand-new John Deere tractor that was meant to be following it round the field on the first cut of the first crop to be harvested. GBH to both, embarrassment to the student, and total calm and dignified behaviour by the farmer!

It was not just the less bright who made the ghastliest opening moves: the double-damage accident was the responsibility of my wife's cousin, Peter, who was one of the best. He obviously took the whole thing to heart, and was discovered by his brother, Tom, climbing out of their bedroom window in his sleep, with his tennis racquet, and saying he had to go to shovel the bin out! Tom, in his turn, swiped through the wooden roof timbers of the 'Pot Shop' with the mast of the fork-lift up. This building was originally the old working-horse stable for a dozen or so shires; after that our seed-potato store and chitting house, and when the potatoes went, home to the Halford collapsible lambing pens. A sore point there, as I entered them in the Good Ideas competition run by the *Farmers Weekly*, only to see them being offered for sale commercially the next winter.

The shed still had its original timber-and-slate roof, that is until Tom Beharrell modified it. If the name rings any bells with golfers it could be because their father, John, had been the youngest-ever amateur champion at the age of eighteen.

Some came back for more again and again, and became extremely knowledgeable and valuable members of the team, not excluding my own boys, who were great to have around. Each of my four in turn took on the duties of the B-team. We sometimes changed the combine in such a way that we had the luxury of two for part of the harvest, and so the B-team, with myself driving the old machine and the boys carting the fringe or more specialist crops back to the old drier set-up at the home base. We had our dramas (see later in the disaster section!) but this allowed Peter and the A-team to get on with the serious business of the wheat harvest.

Christopher Miles tried to get his own back by sending his son, Hughie, to me for a year. This backfired as Hughie was ten times more competent than I had been for his dad, and none of the grievous mistakes quoted above were in his case history. His reign at the farm was more noted for his amazing social life and stamina. It was all the fault of his brother and sister, we were led to believe. Both at Cambridge, and determined that their young brother would not miss out, they ensured that he was invited to tea parties, 'Muckle Flugga'

(Scottish dancing sessions) and anything else going, which had the effect of keeping him out of bed until at the earliest 2 a.m. every night. He was a complete zombie for the first half of the day, and gradually became paler and greyer as his eye sockets receded further into his skull. He timed his alarm clock so finely that he could just make it out to the morning line-up for 7 a.m., but would then perform in his sleep until breakfast. By 'Muckle Flugga' time he was a ball of fire again, but in the end I had to be really cruel and wake him up at 6.30 in order to switch him on.

We all got pretty exhausted through harvest, and longed for the odd wet day; not too wet you understand, but just enough to allow us to catch up with the corn heaps, and mend the odd machine. One, a Kiwi, who worked as a couple with his wife, managed to get a quarter of an hour's shuteye on the kitchen bench after lunch each day; another went to sleep in the bath, and fearing him drowned when I could get no response, I had to climb up a ladder through the bathroom window to extract him before he slipped beneath the waves. The latter, Bill Comerford, now farms in Devon, but his father still lives in Great Gidding. The Kiwis, Lindsey and Glynis, live near Miriam, whose wedding we went to at Invercargill, and we have called on them several times.

Mention any one of the slaves and a picture will flash up. Maybe the one of the grain trailer, loaded but tipped in the field, standing on its tailgate with the drawbar vertical. Neighbour Dick Warrener looked on over the boundary hedge, wetting himself as Sue and Kate struggled to clear up the spillage with dustpan and brush. The same good neighbour watched our combine set fire to the field of standing wheat, and rushed in with all his tackle to save the crop, the combine and the day, while our state-of-the-art radio link failed to negotiate the hill that lay between the field and home. We shared a common boundary for about a mile, and it was amazing how one side of the hedge could have a different climate from the other. It worked both ways: my field was devastated by a hailstorm while Dick's was barely touched but, to make up for that, Dick had to watch our combine rolling merrily on in the sunshine, just after a huge deluge had stopped him for the day.

Peter Horn? Ah yes, he was seen throwing his hat into the combine's mouth after ramming his second electric pole. Jonathan Boldero? Ho-ho! Hopping along as he desperately and unsuccessfully tried to stop his grain trailer from careering into the back of my car; he was all right, and we went to his wedding a couple of years ago, but the car was never the same again.

Only once did we have to say goodbye to one in the middle of the action: it was in fact during haymaking, when Robin was carrying a buckrake load of bales back to the yard. Travelling flat out down the drive, he hit the concrete where it had a slight ramp; the front wheels of the tractor reared up leaving him without any control of direction. Yelling at the top of his voice, he careered into the old hay barn and came to rest with the corner of the building enfolding him and his load. The next day he refused to get on the tractor at all, but I managed to persuade him that tedding the rows still to be baled would be a gentle and stress-free way back. This would have worked had he not driven very slowly up and down the same piece of field all morning, too scared to move on to a new 'land'. At that point we thought he needed a weekend at home to recover his nerve, but he never came back! I really don't think I was horrid to him!

The girls were usually helping with the lambing, or cooking for the team at harvest. The latter was a major feat, and tested the most domesticated of them. Karen, from a farm way out in the bush, 700 miles from Brisbane, solved her problems by means of daily faxes to her mum, who sent recipes and tips in return. Sue and Kate, last heard of sweeping up spilt wheat with dustpan and brush, were working as a pair when my wife was out of action having had a major operation. Normally, though, my wife had small children to look after and the additional help was to keep the other slaves fuelled. There were many who were never filled up, and as long as there was a scrap of food left they would be hungry, and would raid the fridge or larder. We couldn't understand why all the marmalade was disappearing, until we were woken by a crash from the kitchen in the middle of the night, and all was revealed! One poor chap was studying for A levels, having decided to become a doctor at the age of twenty-eight; at the end of the day he had to keep himself awake for another couple of hours' revision, but still managed a bit of hanky-panky with the cook! What a man, (but he did look rather tired, and went grey comparatively young).

Some of the great failures were nothing to do with the students! The farmer himself made some mammoth misjudgements, and grew some of the most disastrous crops ever seen in the district. My efforts at peas were always thwarted by the pigeon population, which I maintained single-handed; embarrassing was not the word for it when my bank manager came out for a day's pigeon shooting and could see for himself what ludicrously optimistic forecasts I was making for the crop.

I was always tempted to try new and more-experimental variations, in particular to find enough break crops to maximise the first wheats; thus it was that I planted fifteen hundredweight of lupin seed in one field and only harvested nine hundredweight back (*in toto*). That was an improvement on the sunflowers (*napraforgo* in Hungarian!), where the birds had taken every single grain in the twenty-four hours between my deciding to cut the crop and taking the combine in, although admittedly only half an acre went west here. I was quietly pleased to learn that my old friend, John, from Cirencester, now farming in Suffolk, had also been experimenting. He had always been a great man for new ideas, and I gathered that his efforts with wallflowers for seed, and onions, had given him his share of problems too!

There were some successes to keep me going; by sheer luck I started growing rape, as a result of establishing a fine stand after early-lifted potatoes. The original theory was to discourage the dreaded wheat-bulb fly from laying its eggs on the bare ground. However, the rape crop looked far too good to plough in, and then far too good for the sheep to graze, and it was left to flower and ripen *au naturel*. By this time it was HUGE — well over seven foot tall, and posing unspeakable problems for the combine driver, but clearly carrying a great weight of valuable seed. Hence a trip to Sweden with Twyford's, and the revelation of what they were achieving at Svalöv where they treated it as a crop in its own right, breeding special oil-seed varieties etc. I imported some of their Victor on my own account, grew one of the first specialist oil crops in the UK, and began a long, happy and profitable relationship growing all sorts of seed for Twyford's.

The advent of rape as a viable crop did at any rate give me the chance to wave goodbye to the potatoes, which had been responsible for many a nervous breakdown: if I managed to plant them in the spring I would find them too wet and muddy to sell in the autumn, and another nightmare that comes back to haunt me is the potato merchant, ringing from as far away as possible (e.g. Cardiff or Bristol) with the information that he couldn't sell my load and would I like them back or take £5 a ton for them. About one in five crops made a healthy profit; the same made an enormous loss and the other three just about broke even. We had a great gang of local ladies hand picking them, and I still get hailed happily in Oundle and nearby villages, by many of them who had enjoyed the fun and general atmosphere of busyness in the spud fields. It still came back to the basic fact that my first judgement was correct: it was not root land, or anywhere near it!

I have to admit to having been pretty successful with beans (which replaced the peas but without the pigeons) and with linseed, which looked so beautiful when it flowered that it was almost worth growing just for that. Add to that the acreage payment of £200+, and we had another profitable break crop. All the while this experimental farming

C

was going on, the wheat (my banker) marched onwards and upwards; massive inputs of research from the plant breeders and the advent of dedicated agronomists, as well as researchers in Belgium and Schleswig-Holstein, allowed the crop no rest! From the two tons per acre of the new wonder-wheat (Cappelle Desprez, in 1957, a fantastic milestone at Christopher Miles', which had caused me such grief on the bagger combine) to my last harvest year in 1996, when the best crop yielded over five tons per acre, and the wheat averaged eighty-four hundredweight, was an amazing increase over forty years.

It was David Boothroyd's bad luck to be appointed to be one of Cambridge Farmers' agronomists after the record harvest of 1984, when I joined. That year beat all previous crops by miles, and it was a long time before he was able to match it. He did it in the end, to cheers all round, and in the thirteen years that he was my advisor, until I retired, he was a great asset to me and all his flock. He strode the far horizons with seven-league boots, and few could keep up with him as he cross-pollinated ideas throughout his group. He came to know our farms better than we did ourselves, and kept us ahead of the opposition in using the latest technology and treatments. Ever the optimist, confident and encouraging, he was so complimentary about us and our farms that we always looked forward to his weekly visits as a welcome boost to morale. He was and still is a very nice man and a good friend.

Flying Fingers!

Chapter 5

Over the Hedge

This can cover anything, but for those who may be worried about my indiscretions, it will make a change from boasting about my winter-wheat yields, and there will be nothing to bring a blush to the most sensitive cheek! To set their minds at rest from the outset, I had better say a bit about my farming neighbours, who have not, as yet, paid me anything to influence the content of this section! I was extremely fortunate in those who surrounded me, and the next ring beyond them and so on. As all our acreages got bigger, they tended to be further away, but we all begged and borrowed from each other, as before.

I have already mentioned Dick Warrener laughing at our predicament when the grain trailer stood on its tailgate, but he was very quick to help out at any time when needed, and a good friend for many years. When I took over Grange Farm, my predecessor, Harry Berry, introduced me to Dick's father Bill, as "the best neighbour anyone could have", and it has run in the family. Bill passed on in the fullness of time and Dick is now handing over to his son Geoffrey. However, he still drives the combine, likes to play with the digger every winter, and has the best-maintained ditches for miles. He also used to come and shoot at Hamerton, and I have seen him bring everything down; but then to stop any tendency to big-headedness, there was one drive in the wood where we picked up dozens of empties, with very little to show for it! We do all know that feeling though, don't we? He also took on my very good shepherd, Fred Williamson, when our 1,000 lambing ewes proved to be unprofitable too consistently. Fred had come to me from Sandringham, and during his time with us managed to get all the tractors at Grange Farm, and most of the district as well, tuned into Radio Norfolk.

Moving round in a clockwise direction, across the old Roman Road is the Leighton Bromswold Estate, owned by the Church Commissioners, and the tenant of Salome Lodge Farm was a fiery little Lancastrian, John Thornley. He had made his way up from a small dairy farm near Preston, with twenty-five acres, a few cows and a milk round, to Cheshire, Lincolnshire and finally to Huntingdonshire. There he seemed settled, with a milking herd approaching 300, a hundred acres of sugar beet, and a keen interest in Peterborough Football Club, of which he became chairman. Unfortunately there was some irregularity in paying a player too much at a time when it was strictly governed, and John had to carry the can. Not just because of that, I am sure, but because wider horizons beckoned, from the far side of the world in Western Australia, onwards and upwards he went. He once confided in me that he had no difficulty making money, but was short on ideas how to spend it; what a partnership we could have made, I thought.

I was partly responsible for a mass migration at that time, in that as chairman of the local discussion group at Kimbolton, I had been to Western Australia House to see the agent general and booked him to come to speak to us. His talk was mouth watering, and we had a very impressive turnout of members, but he had forgotten his slides. When he offered to come back with them and show them at our AGM, it was as manna from heaven! The regular attendance at any AGM is pretty pathetic, and our usual expectations were no different; in this instance we would have normally had twelve to twenty people in the Mandeville Hall, but the added attraction of Gerry Wilde brought in 120. As a direct result at least six families emigrated, and met with varying success over there; some are still farming! John is, sadly, not one of them; he bought a very good and well-developed farm in a decent rainfall part of Western Australia. I passed it some years later when I went over to see my daughter Bridget, at university in Perth. The Ponderosa, near Albany, was a delightful property but was not on the scale to satisfy John's ambition, and to make up the acreage he bought several thousand more of semi and totally undeveloped bush. At a time when wheat quotas were just about to come in, and severe drought was to strike, this was a ruinous move, and very bad luck. John was only receiving sixpence a lamb, and was not allowed to plant wheat without a quota on the new land, so had to try to sell it. Finding no buyer, he had to leave the Ponderosa and come back to Lancashire where he ended up as he started, with twenty-five acres. He had no milk round this time, just a few ponies, and he died some ten years ago.

John had provided a lot of amusing topics of conversation, not the least of which was his ongoing feud with his other main neighbour, Bob Eayres. They contrived to fall out over everything, whether it was the dairy cows being let into Bob's crops, spray drift or, in the most spectacular case, a bout of fisticuffs out on the boundary, because one of them had supposedly dug out a ditch the wrong side. The sight of the giant figure of Bob, ducking and weaving as the bantamweight John darted around him in circles, throwing punches (which never reached their target) right and left like a professional in the training ring, was amazing. I did not witness it myself, but it was long spoken of with awe.

We had not said goodbye to the North for long, though, because the new tenant for the farm came from Cumbria, and we had a language barrier once again! Frank Stamper arrived in the autumn of 1968, and immediately set about showing us all how the heavy land should be farmed. In fact, the first time I met him he was ploughing. I walked across the field to meet my new neighbour, and was greeted by a broad grin in a friendly face. All was obviously going to be well! He also had a large milking herd, and was a stockman at heart; when he took a golden handshake to get out of milk, the farmyard became a dead place compared with the throbbing activity before. Frank was utterly miserable without the animals, and was round to his neighbours (us) every day to see what we were doing. It didn't take too long before a few beef cattle crept back on to the farm, and soon there was a substantial herd of pedigree Simmentals. Meanwhile the arable looked better and better under his regime and he had an endearing love of new machinery. It was a great source of joy and speculation to his neighbours, who never knew what wondrous piece of equipment would turn up next. For nearly thirty years he was my closest neighbour, and the one I saw most of; round for coffee most Sunday mornings, and always ready to help out.

He only once tried to kill me! As I walked along my farm track at 7 a.m. with the dogs, I heard a zinging noise over my head, followed by a thump into the field about fifty yards away; a minute later it was repeated, and at that point I thought I had better protest. As it involved a mile walk back to the yard to pick up my Landrover, then a drive over the hill to Frank's, the bird had flown when I got there. His son, Lennie, was most encouraging, saying, "Dad's always doing that, trying to scare off the pigeons." So that was all right then! A more interesting method of pigeon patrol was radio-controlled aeroplanes, but Frank's piloting left much to be desired, and he wrote off the Pigeon Patrol several times. In return for that attempt on my life I nearly gave myself a heart attack when I rushed across the fields in answer to a desperate series of calls for help. After a mammoth sprint and switchback ride in the Landrover, I was somewhat aggrieved to find no human in distress, but one lonely peahen on the barn roof! Frank may have handed over the tenancy to his son, and technically retired, but still puts a full day's work in most of the time; he also pauses to play golf (far too well for me). The third generation is now helping on the farm, and young Robert has passed his driving test; we will all hold our breath and look both ways for a while!

Moving round the boundary, I only had to sniff the breeze to tell me that my next neighbour was still in business. When I arrived at Grange Farm, Bill Hurley was milking a small herd of cows on thirty-odd acres; he was always to be seen walking them to and fro, but not having a lot of scope for expansion, he changed to pigs. "Pigs is different" as we all know, and in no time at all there seemed to be thousands. His son, John, took over the running, but Bill even now does a lot of work on the farm, and also finds time to play golf. He has never looked any older than the day I first met him in 1959, and the moral of this must be that pig shit is good for the health! The prevailing wind brought all the joys straight to us, but apart from that they were always good friends and neighbours. For a while we swapped straw for muck, but we couldn't cope with slurry on our arable fields, and the arrangement died naturally.

Now along the North-Western Frontier, the Khyber Pass of Huntingdonshire, the local warlord is a member of the Tebbit clan, whose present chief almost single-handedly sustains the local saloon, The Black Swan. The family came from Toft, near Cambridge, and still farm there. Derek had moved to Old Weston, and was one of the first to come to say hallo. In fact 'passing the time of day' took on a whole new dimension with him. On one occasion I had my small son, Peter, in the back of the Landrover, having collected a crate of milk. Seeing Derek at his gate, I stopped, as one does, and half an hour later the two of us were surprised to see a river of white come flowing towards our feet from the back end of the vehicle. It was young Peter's way of protesting, and we only had the odd pint left by the time he had finished taking the lids off and pouring the contents on to the road.

He always seemed very laid-back, and the pipe clamped firmly in his teeth added to that look. I called at his house one day, to tell him I had just put one of his ewes up on her feet, braving his fearsome pack of guard dogs to do so. The farm being at a rather exposed road junction, and vulnerable to any passing thief, the pack was chained up at various points round the yard and all access gates. As my car drew up and was mauled by the first sentry, Derek said, "You don't want to worry about Ould Bob — his bark is much worse than his bite." I was quite relieved to hear that, as Ould Bob's bark had punctured the front tyre, right through the side wall, and I hesitate to think what his bite would have done. We watched the car subside to a gentle hiss, and my neighbour looked faintly embarrassed!

Sadly he passed on quite young, and was succeeded by his son, Dan, a younger version. The most noticeable difference was that there was no pipe, and that Dan's knees were shot up by his misspent youth on the rugby field. He is an impressively good shot, and was, I thought, as relaxed as his father until he had a bad heart attack, apparently caused by stress. As I write this he is recovering at home, and it is a good thing that his daughter, Ellie (running the meat-packaging-and-selling operation), has just moved back home with her new husband, Russ. He can now do all the work and the worrying can be divided up. Ellie came to me to learn her lambing, and had a lovely cure for any weak or sick lamb — she just cuddled them better.

It has just struck me that I am being chauvinistic and forgetful in failing to mention the power behind the throne in all these cases. All the way round the boundary fence there is a loyal, hardworking and friendly collection of other halves. So I make a sincere apology to Margaret and Lucy Warrener, Sylvia and Pauline Stamper, Pauline Hurley, Norah and Maggie Tebbit and, in advance of my omissions to come, Gladys and Helen Steel, and Rita, and Ruth Berry. All have been welcoming, hospitable and helpful at all times and made life so much easier.

I had another language barrier on my northern boundary when I first came; Jack Steel's parents had come down from Scotland in-between the wars, and retained strong accents. In Jack's case he could have been speaking Scandiwegian for all I could understand. The first time we met was when he and his sheepdog gathered my newly-delivered Clun ewes from the far corner of his farm and returned them. Once I tuned in to the language we had a good relationship and a weekly appointment at The Green Man for a drink and skittles. He died very suddenly of a stroke, when the eldest of his three boys was only twenty-one, and it was touch and go whether the family would be able to keep the tenancy. We all watched John and Alan farm their socks off, with Gladys's guiding hand, and they never looked back; in fact they put us of the older generation to shame by their work rate, organization, and their standard of husbandry. Now, when the whole industry is suffering from lack of profits and the need to diversify, they have made a very successful business out of bridge building and similar work. It started when they learnt what the Council were paying for some new bridle-path bridges over the Alconbury Brook. 'Outrageous,' they thought. It was a monstrous waste of taxpayers' money, and they said so. When asked if they could do any better, their answer was obvious, and they have been as good as their word. In their turn they have teenage sons helping out at harvest.

Their younger brother was too young to be included in the tenancy at the time, and in any case there was not enough scale to support more, so when he left school, Robbie started a spraying business, which developed into a big and far-flung operation, travelling from Luton to King's Lynn. In fact it seemed to have grown too big for its own good, and the equipment and men spent more time on the road than in the fields. When he got the chance Robbie took on whole farm-management agreements, which is where he is now, at a very tough time for all farmers; the spraying business was sold out to the men who worked for him. Ever cheerful, undoubtedly helped by having a super wife in the champagne business, Robbie can be distinguished at considerable range by his coot-like 'yipping'. Throughout any conversation he will keep up a constant flow of response with, "Yip-yip-yip-yip-!" There is no trace of a Scottish accent in any of the three, but when Robbie and Louise married there were more kilts and knobbly knees on view than we had a right to expect down in southern Huntingdonshire.

Finally round the last bend into the finishing straight, I come to the Berry family. Bill had grown up at Grange Farm, but being the elder of the boys, had been let his own farm when he got married. He and Rita lived in the village, running a farm and the village shop (with off-licence), and were very much the core of village life. Here again we are on the third generation, as son John took over and is now handing over to Marcus in his turn. Bill sadly died some ten years ago, but Rita is still going great guns. John meanwhile has diversified into servicing lawnmowers and other garden equipment; his wife Ruth, came from a farm near Thrapston, and I used to buy my rams off her father, Gordon Cheney. He was a delightful old boy, but came to a tragic end when he fell off his haystack in the barn, and broke his neck. Ruth plays a very strong game of tennis, but her main claim to fame was as a restraining influence (?) and favourite shopping companion of Vicky Cuthbertson, who with Bob and their two boys were our next-door neighbours at Cottage Farm, while we farmed the land round them. It was particularly exciting tipping the grain trailers at the store near their house, as they bred Dobermans; in themselves they looked fiercer than they actually were, but were triggered off by a little brown poodle, and many is the time that the student trailer drivers had to leap into the tractor cab to save themselves. Bob came from Australia originally, and they have now retired there: first of all to Adelaide (too cold); then Cairns in the north of Queensland (too hot!); now back in Adelaide, which seemed lovely to us when we went to visit them.

My increasing acreage, like many others', only came about through another's misfortune: I had lost two of my original neighbours within a few years. Albert Martin at Salome Wood Farm had become seriously ill with a brain tumour, and had to give up at short notice. On the other side my neighbour had got into trouble with greyhounds, amongst other things, and had also failed to take in his harvest. I watched in surprise as all his sacks of corn were left out where they had been dropped in the field, getting wetter and wetter, and no sign of any attempt to cart them in. The birds and rats helped themselves, and in the end even the hired sacks were a write-off, hung out on the barbed wire with their bottoms rotted out. It was no shock when he went bust and disappeared into the night. His attitude to sacks of harvested grain seemed to be rather Irish; this thought only came to me on a family camping holiday in South-West Cork later on — it must have been a foot-and-mouth year, as I was sprayed from head to foot and made to take my shoes off by the customs! The point about the sacks, though, was that when we pitched camp in the field the sacks were lying where they had been dropped by the combine, and obviously had been for some days; I stood them up and waited for the owner to clear them, which didn't happen. I turned them the other way up, shaking out the corners to dry, and waited some more. Same again, and so I turned them once more before we left for home, and never saw anyone show the slightest interest. They may be there now for all I know!

Others who bordered me, but whom I saw little of include 'The Messrs Langley Bros' as Jack Ruston always referred to Derek and Frank; staunch Methodists and very traditional in their farming, we had one field with a common boundary, but did not meet for several years. They turned out to be a delightful household, with Marjorie, and had a wonderful collection of old photographs which gave great colour and authenticity to the millennium history of The Giddings forty years on.

Farther Fields:

Apart from my close contacts with those round about me, I joined several groups, and there were many trips organized by seed firms, ICI, and so on; my visit to Svalöv in Sweden to study rapeseed production is an example of the latter, but I did not mention the visit to the cricket club dance in southern Sweden! It seems unlikely now, but when we stayed at Simishramn we discovered that we had hit the event of the year. I regret to say that our party did its collective best to cut in on young couples and make off with their women. It was all reminiscent of *Seven Brides for Seven Brothers,* and my roommate had to be persuaded not to weave his way drunkenly around the corridors after lights out. Following on from that, my own visit to a friend's family up north of Stockholm was very tame and civilised; in fact I stayed in a stately home called Margretelund, and was royally entertained and shown round by Gregor Aminoff and his family, and had to reply to a welcome speech after dinner in Swedish. It was a short reply, of necessity!

I have mentioned the Kimbolton Discussion Group earlier in relation to the Western Australian migration that I helped to provoke, but that was only one amongst many organized by the KDG. Originally the groups were set up under the National Advisory Service initiative, to bring farmers and farming up to date after the war, and the success or failure very much depended on who the local advisor was; in our case Frank Coles was a brilliant and pro-active organizer and communicator. It was he who was first on my doorstep and recruited me, and also gave me a mass of helpful advice and local contacts. The discussion group met once a month through the winter, for talks from well-known figures, and also ran popular summer visits to farms in another part of the country; which usually lasted three or four days and took us to the Lincolnshire Wolds, Romney Marsh, Scotland and Shropshire for example.

When my turn came to take the chair it had already been decided to visit that well-known farming county, Lancashire. I and a Lancastrian emigrant were deputed to recce the area not covered by Manchester or Liverpool, and so it was that Doug Robinson and I set off for the frozen North. Things became clearer as to why, when our first port of call was Doug's family cotton mill near Bolton; from there we went to the Pilkington glass factory ('Furnace 3000 Degrees C. Do Not Walk' — 'in bare feet,' some wag had added), and another two industrial sites, before we did finally find some farming. In fact it was supremely high-powered vegetable growing, and there was also some very impressive stock. Our most important task was to ensure that we were comfortable and well fed after our exhausting days, and this search found us at Southport, at The Prince of Wales Hotel. Fine so far, but what entertainment was there in the evenings? Even better, we discovered the Kingsway Casino, a gambling den combined with a Bunny Club. Having negotiated a special rate for our party's evening entertainment, we were satisfied with a job well done, and came home triumphantly.

Imagine our dismay when we arrived six weeks later with our eager coach-load of forty-odd members, to find that the Bunny Girls were all on strike, and only the Bunny Mother was on duty. She, as we thought only right and proper, was allocated to the senior (i.e. the oldest) member. Everything else went like clockwork, but the abiding memory in most minds was their disappointment!

Another group that was a great inspiration and incentive to its members, was the Clayland Farmers; this particular one, among a host of like-named parties, was deliberately kept small enough to look at each other's farms, warts, costings and all. We did this over two days in the summer, mocking in turn my potatoes, Gordon Sandercock's sterile brome, Ebbe Heyman's potatoes, and so on. We went through the costings and gross margins in one of the winter sessions, and salutary it was too. Initially John Jenkins and his manager Dick Wreford, swept all before them, and the Childerley Estate always looked wonderful. Of course it was bound to happen that we all played catch-up with each other, copying the best practice, until there was less and less difference in our wheats etcetera; Ebbe and I dropped our potato crops, too hurt by the mocking cries that greeted both the crop and the subsequent figures.

Dick Beeby soldiered on with his 100 acres, but only because his father did the whole lot by himself, and his resulting margins were huge; in fact we tended to discount Dick, as he did everything so well, and whatever he turned his hand to was bound to succeed. It had been athletics in his youth; we would never have known if his dad hadn't said one day, "Did you know he ran in the invitation Emsley Carr mile race at the White City Stadium?" No of course we didn't, but were not surprised to hear it. After all he had just come in the top six of the RAC motor rally, having been champion go-carter before that. He came to farming from another business, building I think, but quickly became the best arable farmer in the area. Only later did he compete at the highest level with his ponies, driving pairs against the Duke of Edinburgh. Sickening how some people have all the talents! I was not totally dismayed when he failed to become world champion at golf, but to make up for that he took up singing at the tender age of seventy, and the next we knew there was this wonderful tenor voice, singing solos and giving concerts. Now he has just remarried after fifteen years of widowhood, and got a star there too. Well done Dick!

This group folded up when the almost universal use of the professional agronomists levelled everyone's systems, yields and general standards of husbandry up to that of the top; in fact it was notable that those who did not join any of the Cambridge Farmers' agronomy groups quickly dropped down the ladder. Although we had our share of David Boothroyd, he in turn was part of a wider group of six, pooling all the information between them, and we got the benefits of this ongoing research from over 50,000 acres.

Amongst other notable trips was one to study flax growing in northern France, where we were particularly impressed by a new processing plant that had to operate at such pressure to justify the capital that one of the men had lost an arm trying to unblock something, only a week previously. The object of the exercise was to set up a flax industry in England, but it never got off the ground, as all EU grant aid seemed to be directed at improving existing enterprises, and in our case there was none at all to start from. There had been a flax industry many years ago, but it had faded out, although many of us were now growing linseed for the seed itself. (It is basically the same plant as flax, but with shorter straw and more seed.) It looks so beautiful in flower, as each morning dawns with a fresh bloom on every stem, and the impression is of a peaceful stretch of water.

We also went to the same area of France and over the border to Gembloux in Belgium, at the time when Professeur Laloux had revolutionised the approach to growing wheat; this was as the Clayland Group, but the thirst for knowledge led us by the nose, and we found some remarkably good restaurants and night spots on the way. Arras was one interesting centre, and not just for the farming, but because my father had been wounded in the First World War nearby, and I still have the map of the trench system from 1916. Arras was also where the battalion headquarters was overrun and captured while Dad was away, causing him to be promoted to company commander on the spot. One farm we visited had been set up some ten years after the end of the war, having been designated as a mass grave. Identifiable items of uniform, etc. were still being retrieved as we visited, and I brought back some military hardware; Dad was not very impressed to be reminded of it. The haunting memory of those old battlefields, with a little cemetery in virtually every field, affected us all deeply, and for me the memorial at Thiepval was the most moving: 78,000 names of those British and Empire soldiers who disappeared in the mud of the Somme battles.

Brussels was a good base to work from! Apart from the wheat, there were some wonderful craft workers in the arcades at night (no sniggering here please). There was also Waterloo, and it was fascinating to find out that we apparently lost that battle! Only the last minute intervention of the Prussians saved us, and the graphic cavalry charge by the French Cuirasseurs had us all turning tail and running back down the hill. How the young British officers managed to fight at all after dancing the night away at the Duchess of Richmond's Ball I could never fathom, when I had difficulty in performing the simplest routine tasks after nightly dances at the Hyde Park Hotel and others during our strenuous tour of duty in London and Windsor.

David Boothroyd took his little flock on several expeditions to improve our ideas, and one which sticks in the memory was that to the Border country of the Berwickshire, Roxburgh, and Hawick area. There the Rennie family had just broken the record barley yield with nearly six tons an acre, and we came away full of admiration for the Scots and their farming. All except one place had three equally valuable enterprises on which their businesses stood firm: there was a strong livestock enterprise, a major root break and the combinable crops. The exception to the rule had a young go-getting farm manager who was going to set the world alight; he got rid of the livestock, in that case sheep, "because the buildings and handling facilities needed a lot of money to bring them up to date". With the sale results he launched into high-level cropping of spring rape and even wheat, up to 1,000 feet above sea level, where the sheep had grazed previously. To cater for the enormous crop anticipated, he had built a state-of-the-art grain-handling plant, huge drier, weighbridge and all. We gawped at it all, but were not in the least surprised to read about the ruination of the Marchmont Estate and its subsequent sale out of the McEwen family who had owned it for hundreds of years. It was very sad, entirely predictable and unnecessary — if only someone had said no to the whizz kid.

Judging farms and crops in other parts of the country was also quite sobering; everything was so good! I started with my basic scorecard, but quickly discovered it to be nearly useless as everything seemed to be getting full marks, and I had to formulate my own variations to provide some differential between farms. Matters were not made any easier when I found that, as a wheat grower who had let it be known that he was completely ignorant of malting-barley growing techniques, and therefore anticipating a role as judge of the milling-wheat section of the South Suffolk Crop Competition — guess what, no prizes! I fear that the spring-malting-barley growers had a raw deal from their judge that year!

The Newbury District Agricultural Society Farms Competition, over-2,000-acres class, was a revelation; the six or so entries had to have a half-day visit each, and the standard of farming was staggeringly high. Blocks of wheat (up to 400 acres in one case), which were so thick that a cat could walk across the top, were the norm, and we had to separate them. As it was a farms competition, not just crops, I and Frank Stamper, my fellow judge, devised our own scorecard. It involved such things as headland management, environmental features and habitat variation, quality of individual enterprise managers and motivation of staff, cleanness and happiness of the pigs, and adequacy of the stand-by electricity supply. The workshop facilities and grain handling were included as well, of course, but at the end of all that we still found very little to choose between Linkenholt, Littlecote,

Longparish and the Hollands near Hungerford. There were two others in the class, who were not far behind, but in the end we decided that Robert Bowden at Longparish had the best-looking secretary! He also had an aircraft carrier engine as his stand-by generator, and his hedgerow management with its grass buffer zone was impressive. The others had spectacular features about them, and the shooting at Linkenholt, and the sheep enterprise and new shed, were both in a class of their own. Andrew Lloyd Webber's estate had THE Watership Down as its star piece of environment, and altogether it was a great education in what a wonderful industry and countryside we have around us.

This was all some fifteen years ago, in the late 80s, and was brought back to me at the Worshipful Company of Farmers harvest festival lunch, only this month (Nov 2003) held in the Butchers Hall after a wonderful service in the church of St Bartholomew the Great. I found myself sitting opposite Robert Bowden, whom I had not seen since those days long ago. He had left Longparish and now farmed a group of farms including his own, around Dummer. That rang bells with me, and I commented that it was Ferguson country, and that Ronnie Ferguson had gone to his grave owing me £5 which I had lent him in an hour of need in Amsterdam one evening in our misspent youth. Robert reached for his wallet, and with a broad grin said, "I am his executor, and his widow is a good friend of mine! I insist on discharging the debt!" To the amazement and curiosity of all around, he then passed across a £10 note, which I honourably put into the farm cart (our charity collection box) on its rounds at the time.

That was not the end of the coincidences for the day, as it transpired that the Longparish farm that we had adjudged the winner of its class was now, under its new management, looked after by my old friend and agronomist, David Boothroyd, who travels down to Hampshire every Thursday to walk the farm and give advice. My next-door neighbour at lunch turned out to have been short-listed for Frank Stamper's farm in 1968, as had I, and we compared notes on our respective interviews with Colonel Norton Fagge, who was the senior partner of Smiths Gore at Peterborough, responsible for the Church Commissioners Estate, and came down in a chauffeur-driven Rolls-Royce. We even remembered our bid figures: Merlin Usher-Smith had put in £14 an acre, and I £1 more. It would have been quite wrong to let it to me, and I'm glad that the colonel did not.

Chapter 6

The TFA

While all the humble tenants were getting on with their business, it was a time of change on the land-owning front. Our nice Irish landlord passed the estate over to his grandchildren, who immediately sold it on to a pension fund. The first buyers were known as the PFPUTs (Pension Fund Property Unit Trust), and their management committee was very anxious to do the right thing and behave as traditional landowners. To do them justice, they were pretty good at it, and this benevolent group of senior city gents came round on a grand tour every year for a ride on the trailer, a lunch with one of us, and a cheering word. Unfortunately they were not so good at managing the pension funds, and got into serious trouble; the land management side became more and more commercial, and with a couple more changes of ownership we were being squeezed until the pips were definitely squeaking. My rent doubled every three years, and however much I protested it seemed there was little I could do about it.

At the same time as this pressure was being exerted on all tenant farmers, landowners were increasingly reluctant to let farms at all. Legislation in 1976 had slipped in the right for a tenant's son and grandson to succeed, and this could lock up the holding for a hundred years given a youngish family man as the first tenant, allied to reasonable longevity down the generations. There are certainly men farming now because of it, but there was a real possibility that the whole system would grind to a halt, and that no new entrants would ever get a chance. I had my share of luck in getting started, but that was twenty years earlier, and heaven knows how I would have fared in 1979 as opposed to 1959. The other general effect was to increase the pressure on sitting tenants, as they realised that falling out with their landlord would leave them nowhere to go.

That was the background to the formation of the Tenant Farmers'

Association, and a brave gang of four or five set off round the country to spread the word that there would be an organization to look after tenant's interests, and raise the profile of the whole question. This only happened after repeated attempts to persuade the NFU to set up a tenancy section had failed, and was an ongoing source of irritation to the latter for ever afterwards.

I saw an advertisement in the local paper, for a meeting to be held at the Great Northern Hotel, Peterborough, which was hardly an agricultural area in itself; I found myself one of half a dozen in the audience, listening to Stephen Hart and Richard Butler (the other Richard Butler, not the president of the NFU!). Already persuaded before I even went to the meeting, I joined straight away, and found myself in no time a member of the Eastern Region Committee. It was chaired by a Suffolk man, Bruce Seaman, but lacked volunteers for vice-chairman, so of course . . . . It was only a few months after that, when Bruce bought his farm and other smaller holdings with it, becoming at a stroke a brutal landlord instead of a humble tenant, that the realization hit me that vice-chairman really meant chairman.

I found myself responsible for a huge area some hundred miles in each direction, but singularly short on the membership front, with less than a hundred signed up, and a very inadequate committee in terms of their number and geographical spread. The quality was there and there were some high-fliers and industry leaders, but whole areas of desert land with not a single dot on the map, and the national figures didn't have the time to go hoofing around recruiting. There were a lot of big estates in Norfolk and Suffolk particularly, and if I found my way into one of the tenants I could get access to the whole list, so I gradually got to know a large part of East Anglia very well. That didn't automatically bring a full harvest of new members, as some were quite happy as they were, thank you, and had a very good relationship with the estate owner and his agent, which they didn't want to upset. Others were not so happy, and the practice had grown up for the resident agent to be allowed the day-to-day running and general PR work, but that every three years a gun-toting national firm would be brought in to up the rent and put the squeeze on generally. This was meant to keep relationships on the ground sweet, whilst the drain on the farm profits was getting ever worse. Apart from the rents, there were other major problems for those with intensively run commercial shoots over their land — pheasants love young sugar beet! There were heart-rending tales of sons who had branched out from the farm and found themselves debarred from succeeding to the tenancy by virtue of their successful sideline. I was also the recipient of tales of woe of a more personal nature: domestic upheavals, honey traps and unfair prosecutions, depression and quite a few suicides.

Anyway, after a year spent largely on the road like a commercial traveller, I had a membership of nearly 300, a brilliant and active committee, who met regularly at Nigel Rush's farm between Thetford and Bury St Edmunds; an ex-ADAS secretary, Charles Beresford-Knox, who looked like a country vicar (and was in fact the son of one!) was, in spite of his mild and gentlemanly manner, persuasive and efficient. He was followed after a few years, by Mike Trendell, a very active and well-known ADAS man from Norwich, and we organized farm walks and meetings all round our patch.

They were well attended and particularly valuable to anyone suffering the same problems: we looked at Paul Lory's troubles and solutions when the M25 cut through his farm. Peter Faulkner was the valuer who had become a national expert on compensation for that, and had in fact been responsible for getting an adequate bridge over the motorway for Paul, whose farm would otherwise have become two individual smaller units. Peter became a regular speaker for the TFA on many subjects, later going on to head the RICS. Our network of specialists, some fifty strong, could and did handle virtually any problem, from game damage to dilapidations, from succession to rent arbitrations, and it would be difficult to do justice to all if I carry on mentioning them by name. As a result of all this activity, we were definitely being noticed!

At one CLA meeting I stood in for one of our national leaders who was ill, and found myself talking to a packed hall of worried landowners and their agents, but was in the embarrassing position of having lost my own voice almost completely. To continuous shouts of "Speak up!" I tried to put over the main message that we were not anti-landlord, but pro-tenant, and that if we didn't do something about it there would be no tenant farmers left.

The national leadership was in the hands of the giants who had set the ball rolling, and Dick Whittle, Henry Fell, Jim Harrison, Stephen Hart and others had made a huge impact, and were right in there with the chief officials at the ministry, and the various ministers of their day. In fact we were very much part of the top table, with influence far beyond our membership numbers; by then we had over 3,000, but always felt we could have done with more support. Our briefs on tenancy reform were taken to heart, and by the time my time came to sit as national chairman much of the groundwork had been done.

We were quite clearly representing our members' interests far more effectively than were the NFU, who had one or two good men on

D

their tenants' committee, but were on the whole completely ineffective on the ground; so bad in fact that we made a serious push to take over their tenancy role, which we could have perfectly well coped with. Our organization was tight and focused; we had a brilliant brain as policy director — Jeremy Moody — who ran rings around virtually everyone else in the industry, and was able to cut through all the guff in government consultation documents, etc. with no trouble at all. Incidentally his wife, Tricia, is equally bright, and their offspring awesomely intelligent. Oliver, then aged three corrected me kindly when I commented on his model train: "That's not a train, it's a locomotive." — Hmm, quite so. Our network of high-grade agents was proving to be extremely effective in handling tenancy problems, and we had regional and national committees who contained many of the leaders of the industry. Our overtures to the NFU to this end came to nothing, though, as they found the idea of abdicating a section of their territory a complete anathema. The duplication of effort and resources continued, but we did find our views and briefing notes were given high priority by the ministers and senior civil servants.

In my time as chairman I had regular meetings with, in their turn, Michael Jopling, John Gummer and Gillian Shephard. My photograph of the ceremonial handshake between the minister, Hugh Duberly of the CLA, and David Naish (NFU), reminds me of those heady days; the corridors of power are stressful at times, but really exciting and noticeable by their absence when you move away. It was under the last of those ministers that the above photo opportunity took place, and we finally saw an act passed, to free up tenancies and escape the slow strangulation that was the result of too many restrictions on the would-be letters of land. Of course the result was not perfect, and we expected that there would be the need to re-tune the act at some later date. It was unfortunate that it should have coincided with a time of extremely high profits in the arable sector, the effect of which was to encourage wild bidding for extra land at rents (for the new short terms) that became the norm, putting the new entrant out of the running before he had started.

It ceased to be all up to me, as my term of office finished at almost exactly the same time as my mother died, and I had a hip replacement; so although I remained on the executive as a councillor for another eight years, I was not in the driving seat during the BSE crisis, thank goodness, nor the foot-and-mouth disaster that followed on its heels. In fact I retired from the farm at the perfectly respectable age of sixty-five, in 1996, and it was sobering how quickly I became out of touch and past my sell-by date. I had clocked up an amazing mileage (over

35,000 by the end) as I visited the regions, speaking at meetings, attending farm walks, etc. I met most of the membership, and received a huge amount of hospitality and goodwill.

Since my time Reg Haydon has coped magnificently with all that fate has thrown at the industry, and thank goodness for his efforts. He has maintained a high profile both for the organization and personally, and at times has seemed to have a permanent slot on the morning radio. Whereas most of his predecessors had to get back home and run their farms after two years in office, Reg has been able to soldier on for over five, secure in the knowledge that his son is coping very well on the ranch. The resulting deep knowledge of Whitehall and all its personnel has given him many short cuts. Not the least of his assets is his loud voice! Known affectionately in Sussex as 'Foghorn' Haydon, he has bellowed the cause of the tenant farmer around Europe, and one nice little story of him on the Arundel shoot illustrates the power of the human voice. In the steep and deep valleys of West Sussex, which meander lazily along the bottom of the downs, the guns were all lined up. As usual it is only possible to see one or two of your neighbours in the line, and sometimes none at all. However they could all hear the stentorian bellow of our Reg as he sought to whisper some secret information with his landlord, the Duke of Norfolk: "I'D LIKE TO HAVE A WORD WITH YOU IN CONFIDENCE, YOUR GRACE." Of course the whole world stopped to listen, as guns, pickers-up and beaters waited with baited breath for the top-secret details to be broadcast.

This is all a bit reminiscent of an earlier prime minister, talking to one of his constituents in Scotland so loudly that others made the comment, "Why doesn't he use the telephone like everyone else?"

At a time when all farming organizations are losing members fast, largely because people are giving up, the TFA has decided to streamline itself and cut costs by doing away with the network of regional secretaries; for years this battle between central control of secretarial time, and motoring expenses vis-à-vis telephone calls, has rumbled on. A previous director-general, Norris Forster, waged a fierce war with Charles Robshaw of the East Midlands Region, over his travelling expenses, but Charles delivered the goods, kept his membership in touch, and for my money was streets ahead of the other school who did all their calling by telephone. The distances were vast, but there is no substitute for the man/woman on the ground, and to have a situation where members don't even recognize their regional secretary is asking for disaster in my view. We shall see! The personnel have changed, with Jeremy Moody moving on to the Central Association of

Agricultural Valuers, and ourselves poaching George Dunn from the CLA; George has managed to bring the flair and brain-power of Jeremy and the organization of Norris Foster into the same head, and has been a major asset to the TFA.

I made some very good friends through my TFA activities; in the Regional Committee I found myself alongside John Sutton, from Holt, ex-Coldstream Guards officer, and well-known artist, who, with his wife, Carola, showed me great kindness and hospitality on my trips to the east, and opened up dozens of new contacts and members. Nigel Rush and Sarah (with her delicious cakes at every committee meeting) were conveniently central to East Anglia and a tower of strength to the organization; it was Nigel whom I found shooting moles on one dawn visit. I would never have guessed what was going on by myself, as this figure stalked around in the morning mist apparently taking pot shots at spiders, or maybe pink elephants for all I knew. When he wasn't doing these strange things he formed and chaired the Policy Committee — the think-tank of the TFA for many years. John Young from near Hunstanton was another early star, already an internationally known Hereford breeder and judge, as well as a regular writer and broadcaster on the national scene. Henry Cross and his brothers, all from the Sandringham area or the Royal Estate itself, were loyal and active supporters from the word go, and it was Henry who invited me to shoot at Sandringham, and was therefore really the cause of my blowing crumbs all over Prince Philip. The list is endless, and I must not fall into the trap of trying to mention all by name, only to omit key members of the team.

Nationally it was the same, and I found myself in the very fortunate position of having a welcome in virtually every county in the land. I must single out my predecessor as chairman, Derek Jenkinson from Faringdon, as a particular help as he saw me into the job of national chairman so painlessly, having put in an extra year in office, and has remained a good friend for twenty years at least. We have regularly shot with each other, although I could never compete with the lunch room at Buscott Park, where larger-than-life murals of the land-owning family clad in see-through nothings adorned the walls. I did not blow cake crumbs all over my host there, but had the embarrassment of a jammed gun during the best drive of the day. In the 'pound seat' in the middle of the line my gun's ejectors failed, and the extractors got round the wrong side of the spent cartridge: as I desperately dismantled the gun with my boy scout knife, the birds streamed overhead, and my neighbours had a field day. I blame my own ambition for the trouble; thinking to improve my performance on the high birds, I had invested in some expensive cartridges with a name like High Pheasant which proved too much for my old Churchill. The cartridges expanded under the pressure of firing, forming such a tight seal that it was as much as I could do to lever them out with the said boy scout knife. Funny how much business was done on the shooting field!

In the intervening period, my own farm had changed ownership yet again, and I finished my time as a tenant of Corpus Christi College, Cambridge. It was a happy time for me in terms of landlord/tenant relationships; both the college bursars and masters were a delight to deal with, and I really looked forward to the annual lunch in college, with its specially brewed 'Audit Ale'. Once again under the management wing of Bidwells, I have nothing but good memories of my working and personal relationships there: they were straightforward and fair, and I believe we all put in our best to stick to our bargains or go one better. It was a nice change after some of the less scrupulous teams that had trampled over me in previous years.

Lambing time — MJH and Peter Hall.

Chapter 7

The Family Business

At some point in the early 60s I was asked to join the board of the family firm, George Monro Ltd., of Covent Garden. This was not because of any skills I possessed in the field of fruit-and-veg wholesaling, tempting though it is to assume so.

No, it was as one of the few members of the next generation down, and the Monro family (my mother's) needed to ensure some sort of succession. They had been fairly abstemious in reproductive terms, after a strong start under the iron leadership of first of all, old George, the founder, and then his three sons, Eddie, George and Bertie, my grandfather. George (1st) had started his career at the Duke of Buccleuch's in Scotland, where his father was the factor. Promoted down south to Boughton House as head gardener, he had eventually migrated into the North London suburbia of South Herts, where everyone employed a Scottish gardener, and most them from the Clan Monro — making the tracking back for the family records very testing. He wheeled his own cucumbers into the local market, and the business had grown from a weekly barrowload of cucumbers to one of the 'Big Three' firms in Covent Garden.

They were in all the main wholesale produce markets around the country, and in every horticultural growing area as well, where they supplied growers' sundries of all sorts, from string to seed trays to potato bags, from 'Tompots' to the Monro Tiller — the latter was the Rolls-Royce of market garden tractors, and could still be found working thirty years after production ceased.

I found a boardroom with seething undercurrents, and secret family feuds; resentment between different branches was very evident as we discussed each little kingdom within the empire every month. The microscope focused in turn on the enormous proportion of waste produce that was left unsold and rotting at the markets, and the minimal return on a huge turnover of millions of pounds; this was Kenneth Monro's department. (He was the son of Eddie, the eldest of the second generation, who had shot in the national rifle team amongst other achievements.) Kenneth's grilling finished, but not until the small matter of his domestic staff and gardener had been opened up yet again. Why were they on the books? What was his wife's car doing there? etc.

There was a perceptible change of atmosphere as the next victim came in for scrutiny: this would probably be Donald's business, the horticultural sundries, a success on the face of it, but once again having the problem of minimal return on capital, and a huge amount of stock carried through from one year to the next. Donald was at least steady and transparent in his dealings, and did bring in outside people to try to improve matters, but it didn't make things much better. In addition to these problems, Donald and his wife had adopted twin girls (having no issue of their own). They had proved to be more than a handful, both being wild beyond belief, and one committing suicide. Donald got his light relief from racing cars and hill climbs, not on foot!

And so on to the flower department, the only one to trade really profitably, and run by the ex-husband of Donald's sister. Basil Unite had been married (but divorced) into the family but, far from being chucked out, he was much more popular than his Monro wife. What was more he ran a very tight ship, the best part of the whole company, and very definitely the market leader. Where was the family successor to Bertie, my grandfather, you may well ask; after all he had started the flower side, and why hadn't his son taken it on? The simple answer is that Ian had died of TB at the end of the war, having been a squadron leader in Bomber Command, but that is too straightforward!

As an apprentice and trainee he had been sent to a contact in France to learn the other side of the business, and during his stay had managed to convince the managing director of a shipyard that he, Ian, was *the* Monro of George Monro of Covent Garden and wanted two banana boats. This order and its subsequent cancellation upset his father and the other seniors, and cost a lot of money in cancellation fees. Uncle Ian was shot out of the firm, cut out of the will and left almost penniless. He had been quite a playboy in his time, and claimed that he used the same tailor as the Prince of Wales, and had played golf with him! — a case of *I've danced with a girl who danced with a man who danced with a girl who danced with the Prince of Wales*, but it didn't save him from the chop when the time came. So that left my mother, who had the skills and knowledge to run the flower business very effectively, but was a woman! Ladies just didn't do that sort of thing in those days, but she was on good terms with Basil Unite, and her input into the business was through him.

She had to wait for the war to show her ability in management; she ran the local forces canteen which catered for whatever regiment was stationed at Woldingham, and really came into her own. Her accounts from then show what a huge enterprise it was, with hundreds of cups of tea and sandwiches each day, and a halfpenny on each being all the difference between a thumpingly obscene profit and a disastrous loss. She was in the kitchen, fetching the supplies, organizing the rotas and taking more than her share of them, as well as doing the accounts. What an asset she could have been in the family firm! Her sister, my Aunt Peggy, became a doctor, and that was pretty revolutionary at the time.

The memory of Bertie (BJ) lived on, and I found a terrific fund of respect and affection for my late grandfather, who had died of cancer before the war. Although he was the third of the original trio, he was in many ways the most successful and had left his mark on everything he touched.

He also had the gift of enjoying life to the full, and his interests ranged from sailing to golf, and from civic duties to bridge. His sailing may have started in small boats, but by the time I knew him he had acquired a steam yacht with a permanent captain and crew; we had some happy trips on the SY *Sheila*, in considerable style, as she was well equipped with staterooms and all the comforts. She was commandeered for minesweeping after she had helped in the Dunkirk evacuation, and was sunk by a mine; all this happened after BJ had died in the summer of 1939, and she had a new owner.

His golf, played to scratch at Hendon and also St Andrews, seemed unlikely for a man of his size and girth. At nearly eighteen stone it should have been hard for him to see the ball at all, but he could still thwack it straight down the fairway like a rifle bullet. His bridge was meant to be for fun and relaxation, but my mother always remembered his reaction when her innocent childish comment from behind his shoulder caused a change of gear: "I always think the queen of hearts has such a nice face," was not perhaps the best thing to say to a player as he studied his hand, but she was very young, and never took up the game as a result.

He became a town councillor, and was the charter mayor of Hendon when it became a borough. In spite of a full working day which began at about 3 a.m. when he got up to go into the market, he was extremely active in his civic role, and was ably and effectively backed by my grandmother (a remarkable and lovely person in her own right, who

56

was one of the earliest female magistrates).

I mentioned water polo, and though I never watched him he played for Scotland, but against whom I have no idea. Was it an international sport then — in the Olympics, perhaps? I did see him dive off the topmost board at the large Hendon municipal pool — and not just from the board itself, but standing on the guard rail around it. As he stood arms outstretched for what seemed like an hour, we all held our breath, until he did a graceful swallow dive, plunging in with scarcely a ripple to show where he had gone. During his life he had earned life-saving medals from the Royal Humane Society for two separate rescues.

Although he was already ill from when most of my memories of him come, I have vivid pictures of this fun-loving giant; Christmas at Rydal Mount, Hendon, was an exciting time for the grandchildren, my sister and me. Christmas-pud time, and we knew what was going to happen! We would get silver sixpences and maybe bigger coins, but 'Banda' would start coughing and spluttering, and from his mouthful of masticated pudding he would draw out, link by link, his gold watch chain, followed by a final convulsive heave as the massive chiming gold pocket watch emerged. "Really, Ethel, this is too much; sixpences are one thing, but you might kill someone putting this sort of rubbish in!"

This could have been the end of the entertainment, but there was the Stilton to finish; over-ripe and heaving with maggots as our hero mashed them up to spread them on his biscuits, it was not a sight for weaklings. Our parents felt quite sick by now, but the young innocents' joy knew no bounds. A good training for life, I think, although I have never attempted to emulate Grandfather.

He lived his life to the full, and enjoyed the wealth that his business success brought, not slow to share his good fortune with others. One last story about him will have to do. At some big dinner-dance at the Savoy, possibly on New Year's Eve, the supply of waiters in the dining room started to dry up, and soon there were none to bring sustenance to the starving and drought-stressed diners. A recce party was despatched to ascertain whether there was a fire, a strike, or both, but the explanation was simpler. A substantial figure was standing on a table, conducting the entire kitchen staff in a sing-song — *The More We Are Together The Merrier We'll be!* and others. As drinks had been set up by the culprit (BJM of course), it took some time for order to be restored.

My last memories of him are from a final stay at his seaside home at Hythe, where he had a sweet Scot, Nurse McLean, helping my grandmother look after him. He still worked at his desk there, and I used to stand by him in annoyingly angelic silence, until he gave me sixpence to go away and buy a Dinky Toy. Mercenary little beast that I was, it did not strike me as immoral or wrong in any way, until I had the error of my ways pointed out by my mother. I still feel bad about it though.

Where did I come in when I joined the board of this family company? As a non-executive director, I tried to keep the peace, and prevent the whole business destroying itself. I also endeavoured to restrain the empire-building and the losses, but I was at a disadvantage owing to my lack of years compared to my cousins. After all they had been there, tried that, and had all been to war and come home covered with medals and glory, and to say I was there to look after the interests of my side of the family was merely paying lip service.

The chance to get out of all this unprofitable trading came with the approach of a white knight in the shape of Fyffes, known as banana importers, but with designs on a wider field. The deal was done, and George Monro of Covent Garden carried on under new management. Fyffes-Monro continued, but so did GM Properties, having risen like a phoenix from the ashes. We had retained ownership of all the sites, including the three main buildings in London. There was also an attractive portfolio of potentially extremely good light industrial estates at the depots scattered round the country, and a new era as property developers beckoned. Converting and modernising the London buildings was difficult, as we were forbidden to change the appearance of the exterior in any way, but we managed and reaped the rewards.

We had an interesting collection of tenants: in 41 King Street, once the home of the National Sporting Club (before our day), where gentlemen dined whilst watching and indulging in bouts of fisticuffs, we now had the Dance Centre on the upper floors. This was a very exciting tour of inspection for the landowners' party, as major ballet and other dance groups used the facilities, and their dancing was most exotic and unrestrained. The penthouse flat on the top was also a revelation, and we marvelled at the man who suspended his large double bed from the ceiling by ropes, with the bedside tables likewise; on the tables stood an empty bottle of champagne or two, and lipstick-marked glasses. How decadent, how delightful! We envied!

A more conventional but still quite interesting French restaurant on the ground floor sandwiched us, who kept our head office between the two. As far as I know we never got any special treatment or concessions out of either. The same applied to Poons, also tenants, who had the reputation for the best Chinese food in the City, and Joe

Allens (American restaurant and very good too), in the old flower building at Tavistock Street, always known as The Battleship because of its shape. Our old boardroom, panelled and beautifully furnished along with the directors' dining room and loyal staff, were all casualties of this change of emphasis; they were all in The Battleship, alas!

The future looked good for us commercially, but there was the standard family self-destruct button, which was bound to be pressed one day. With only two male members of my generation in the firm, the third down from the founder, we were heavily dependent on harmony prevailing, and that was expecting too much. As with my Uncle Ian and the banana boats, something came up between Kenneth Monro and his son Paul, which rent the temple in twain. I never got to the bottom of it, even though I should have been able to as vice-chairman. The result was that yet another Monro only son was sacked and disinherited, leaving a large proportion of elderly family shareholders wanting their money out, and unwilling to give us the chance to run the show (or abscond with the loot). There were of course some outside directors, and Donald in particular had taken care to put in a good sensible accountant to look after his family's interests. Peter Williams was a very nice, straight chap and the three of us, Paul Monro, Peter and I could have worked together and made a success of the developments. However, when the crunch time came, all the votes were with the older family shareholders, and the firm was for sale.

By then this potentially major property company was wooed by several bigger fish, and finally taken over for some seven million pounds by PHIT (Property Holding and Investment Trust). In their turn PHIT was taken over by Trafalgar House (who in their turn etcetera . . .). A long way from a wheelbarrow load of cucumbers in the local produce market!

The name of George Monro lived on in the Market, long since moved south of the Thames to Nine Elms, first as a subsidiary of Fyffes, and latterly as Page Monro; however I learnt only this week, in August 2003, that the firm had been closed down for good. Unprofitable trading for years had finally been nailed into its coffin by the tidal wave of selling direct to the supermarkets. *Sic transit gloria mundi*, as the old-style Covent Garden porters would certainly not have put it!

Chapter 8

More Farming

With the sale went my second job, which had proved an ideal counterpart to the day-to-day farming problems, and I was going to miss my contact with another world. However, when one door closes . . . and at about the same time I found myself invited along to lunch with the board of the Agricultural Mortgage Corporation. Along with two or three others I was treated to a full sit-down meal of the highest quality, and that included the very decent claret and port with which our tongues were to be loosened. In short, we were expected to sing for our supper (in this case lunch!) which merely involved opening our big mouths and being as frank and indiscreet about the future of agriculture as we saw it. I don't think any of us were unduly shy or reticent under the influence of all this hospitality, and towards the end of the proceedings my neighbour turned to me and said; "Could you do with another 600 acres?" Try me, I thought, and jumped at it.

The question came from the same man who had taken a leap of faith and let me Grange Farm some twenty years earlier, and once again I had to thank Sir Francis Pemberton for my farming opportunities. This new offer came at a singularly appropriate time for me; not only had I lost my outside interest and needed a new challenge, but my son James was about to leave the RAC to come farming; although he planned to work in Australia and New Zealand first, he would need a slot to call his own one day, and I wasn't ready to be pushed out yet.

The final and deciding factor was that a combination of deaths in the family, maturing endowment policies and a payout from the sale of GM Properties, had found me with cash in the bank for once. The Inland Revenue was deeply suspicious of this, and tried hard to link me with some fraud, drug dealing, money laundering and tax evasion, and was most persistent in their pursuit. So much so, that at a final confrontation with them my accountant said quite unrepeatable things to the chief inspector. I sat in embarrassed silence, horrified at John Sterland's approach and strong language, but as each sum of money was accounted for honestly, by the death of an aunt, father, life policy or whatever, and the point punched home with full supporting documents, the inspector wriggled and squirmed and looked all ways for the exit. There wasn't one, and in the end he had to apologise and humble himself. I wondered how long it would be before he would get his revenge, but I had been a tax commissioner for at least ten years, and any flaws in my accounts were not there as a part of consistent and deliberate policy. The inspector was moved on and I stayed put!

I was in a position financially to set up another unit, without milking the resources from Grange Farm, nor relying on my home staff to do all the work. The new farm, at Biddenham in the loop of the River Ouse, near Bedford, was fascinating. I had never seen such a mess of bad farming! The crops were thin and pathetic, but the weeds were particularly vigorous, and the appearance of them was not improved by being sprayed when they were in flower and double the height of the barley. The resulting harvest, which thankfully was not my responsibility, sat fermenting in the store; an average yield of about one ton per acre of unsaleable rubbish. It would have needed a miracle to do worse than that, and my predecessor really made things easy for me; no question of being a hard act to follow! He was a fine shot, and had played county cricket, I was told, but on the evidence before me he was not the best farmer in the world!

The plus side of this agricultural shambles was the proliferation of wildlife in a suburb of the county town. It was possible to have a very sporting little shoot in the first few years, as the River Ouse, with its adjacent meadows, provided a ring fence from domestic invaders, and we had wild populations of ducks, partridges and pheasants. All bird life was helped enormously by the scruffy state of the arable land, and with three or four friends we had very satisfactory bags of twenty to thirty birds in a morning. I suppose it was the inevitable result of the more timely cultivations and cleaner crops that we sought to achieve, but the decline in wild game was soon apparent, and the final straw was the massive fox population. Our little coveys of partridges were stalked as they jugged at night, and gradually were reduced to zero.

When I speak of 'we', this, in the first three years (while James was finishing his Cirencester course and wanderings) was myself and David Hicks-Beech. I had put David in as one-man manager, with most of the tackle that he needed, and he hired his (very!) wild-oat pullers and harvest help locally. We made occasional forays from Grange Farm with major cultivation equipment, but the less travelling with the circus, the better. The roads around Bedford were crowded and tortuous, and the Oakley bridges were impassable to anything wider than a wheelbarrow, causing us do early flits through the outskirts of the town at dawn. David's father, a titled Gloucestershire landowner, was highly amused at this situation, particularly as I was technically the manager (on the agreement), and his son was the manager's manager.

However odd the arrangement may have sounded, it worked well, produced the goods in terms of yields and profit, and carried on from strength to strength under my son when he came home to take over. The expansion of his acreage and business is his story, but one or two memories remain and stick out from the mist of the past.

The whole population of Biddenham seemed to paint their windows at harvest time, and we were unpopular for covering their new paint with smuts when we burnt the straw, and dust when we stopped doing that. One very foggy early morning I went down thinking I would take the opportunity to burn the outside row round a sixty-acre block, to make the fire-break when there would be no chance of it spreading. Nobody was up, the visibility was nil, and I visualised having a real problem getting anything to burn at all. All alone in my eerie silent world, I worked my way round the headland in total peace and tranquillity, dobbing my little flickering forkful as I went. When I got to the end and turned to survey my work I found, to my astonishment, that the whole field had gone up; not a sound had come from all that straw, burnt as clean as a whistle, without disturbing a soul or dispersing the fog. I tiptoed off into another world, where people were awake and the sun shone, very thankful for my luck.

After a few years, with the banning of all straw burning, the problem was much more serious, and we had to contend with a race of pyromaniacs who didn't even wait for the combine to finish before setting fire to the field. We were calling in the fire brigade so regularly that I suspected they thought we were the culprits. By then James had taken on more land, and was farming nearly 1,200 acres, so it added to the problem of stopping the fire lighters. He was often combining on one of the other farms, and so had to make sure that there was someone back at the ranch while there was still any straw at risk. He had two regular men working full-time by then, but it still broke up the party and caused chaos frequently.

Another problem was the fishing line that was abandoned in the meadows; time and again we had to catch and free sheep from this menace, which at its worst could virtually amputate a foot from the poor animal. Add to that the major risk of flystrike in the flock as they grazed by the river, and the luxury of having all that extra autumn grass appeared less attractive. I gave up taking my flock down there altogether in the end, and we let the grazing out locally.

Golf balls did not cause us any trouble in themselves, but did provide a second harvest after the fields were cleared; we were obviously surrounded by dedicated golfers, who were quite prepared to drive off half a dozen practice shots into the fields of wheat or rape, with no hope of retrieval. It must have been rather like practising from the deck of an ocean liner.

Another interesting sideline developed, but not for James. He noticed that there was heavy grazing of the rape crop in the spring, and done in such a regular and methodical way that a new race of thinking rabbits must have evolved. Seeking to report on this phenomenon for science, he went out at dawn and caught a large family band of Asian origin, all crouching in line abreast as they clipped their way across the field. It transpired that they ran a chain of local greengroceries, and that these neatly clipped and packaged bunches of fresh spring greens were much sought after!

He did establish an international airport (for microlights), starting with his own, and a specially-sown landing strip; by the time there were half a dozen regulars based in his shed, plus other casuals dropping in, the good people of Biddenham took exception to the continuous buzz of angry wasps, which was threatening to drown out the sound of their lawnmowing. The ban may have saved James more serious accidents; he had ended one or two promising relationships by landing slightly short of the runway and tipping the girlfriend half out. He had also done a somewhat similar thing to his own father, by landing in the hayfield at Grange Farm and turning across a rather deep tramline. Dad was shaken but not stirred, and James nearly broke his ankle on that occasion.

One of his new blocks of land involved an extremely awkward journey round Bedford and Baldock; it was the latter town that saw his four-wheeled trailer load of seed-corn overtake the towing tractor on the hill down to the town, and smash a combined electricity/ telephone pole. James was somewhat aggrieved to be summonsed and fined for having an unsafe load, but much more so when he received bills from both utility companies for the same pole.

The farm he was heading for was on the northern face of the hills between Royston and Baldock; lovely country, but not without its difficulties. Even on sunny days in the winter, most of the land did not see the sun, owing to the steepness of the slope and the shallow angle of the sun; that meant a slow, late start to the growing season. It was also high enough to provoke any potential rain clouds to drop their loads, and the rainfall was significantly higher than around Bedford. Several times James made abortive journeys over, only to find that it was or had been raining heavily at the other end. Weather reports, from a local station before setting out, became routine.

Combining the rape there, rolling along like a ship at sea over the downs, I felt as if I was the only man on the earth. There were short-eared owls for company, at least ten at one time, and a good-sized herd of fallow deer, but of humans not a trace. Occasionally the fairies would change the trailers over, but I never saw them, being lost in my own world and far away at the other end of a hundred acres. It was bliss! So different from the land of rush-and-tare, blocked elevators and full tipping pits at home!

Sadly for him, at the time, he lost much of his land to development, including the home farm and his main drying building and machinery shed. Having been away from Grange Farm, he of course had no succession rights when I retired, and has ended up with his own farm on South Island, New Zealand. That is his tale to tell, but it is a wonderful country, an excellent farm and a lively and friendly local community; 12,000 miles is a long way to go, but it is a good excuse for travel!

Hungary is nearer, but the difficulties in farming there only became apparent after I had joined a consortium of like-minded optimists, and invested in an arable and dairy enterprise half way between Vienna and Budapest. The language has proved almost impossible for all but two of us, and my sole contribution to any conversation can only be *"Napraforgo!"* Uttered with various accents and different emphasis, it may sound impressive, but still means sunflowers. On balance I am happier with my sunflowers than I would be with another member's carefully coached and rehearsed chat-up line: "Whew, it's hot in here! Shall we go somewhere else?" was what he thought he said, but it turned out to mean, "I'm afraid I'm gay". Oops!

Our takeover of the business at Ikreny alerted us to some of the problems we were about to meet; we had lost twelve tractors and three combines from the original inventory. Yippee! It took some time to track them down to a disused railway yard, and much longer to get most of them back. There is a Hungarian solution to most questions, and it ranges from back-handers, which we don't like and try to avoid, to taking a football team out and allowing it to be beaten by that of the mayor of Ikreny. A likely story, you may well say — we are just bad losers!

The culture of drink was still very strong, to the point where the staff were arriving at 7 a.m., already drunk, and incapable of any useful work. One man drove his tractor and plough into a works bus, nearly killing five people, and our partner/manager out there had to carry a breathalyser with him thereafter. The whole race has an amazing ability to enjoy life, and we found that in trying to hold our annual general meeting in the dining room of the local pub (owned by us!) at nine o'clock on a Sunday morning, we were deafened by the party that had carried on throughout the night, and adjourned to the farm office.

We discovered that we owned the local brothel, probably the most profitable enterprise of the lot, but which we felt ran contrary to our principles! There was also the local store, workshop, and a beautifully run horse business; there were brood-mares and foals; horses at livery and others for hire, as well as riding lessons, etcetera. The place was spotless and the animals glowed and glistened with health. Only one small fly in the ointment cast a shadow of doubt over us; we could find no trace of any income! I don't mean there was no profit, but that there seemed to be no money coming in at all. Of course the farm provided all the feed and paid the wages, so it was naturally a little gold mine — but not for the band of brothers!

The soil is wonderful, deep alluvial silt, but the climate is very different to that at home; temperatures of 40°C in early June kill off most of the crops before they have had time to fill, and for four years out of the six that we have had the business there has been an official drought. To be official the rainfall has to be twenty-five per cent down from the average of the previous four years, so as the average reduces with the succession of droughts, the basis for compensation, if any, reduces sharply. The really desperate drought of the summer of 2003, as I write this, has dried up the Danube to the point when it cannot carry any cargo shipping at all, and only the odd small boat ventures on the trickle that is left. The effect on us is to add enormously to the cost of obtaining soya and other food for the dairy herd, if we can get it at all. A macabre revelation of the Nazi-German Danube and Black Sea Fleet, scuttled to the bottom of the river but now exposed to view and investigation, is another side effect. There are some 200 vessels of different sizes, and this include a hospital ship sunk by her own crew, complete with the patients in her.

We did not envisage the war with Serbia, the destruction of the Danube bridges, the loss of the market for our milk products and the bankruptcy of the dairy company; nor the bill for VAT on the milk cheque that we didn't receive. We did believe most of the

E

encouragement we got from government sources to attract inward investment and, in short, we shouldn't have joined! However we did, and at least were not putting our pension funds into it. It says much for our farming partner's staying power and guts that he has stuck it out; Andrew Hunter's enthusiasm has got us there and kept us going, but I don't see us making our fortunes. It will be nice to think that we can get something out at the end of the day!

In the meantime we provide hours of amusement for our friends with each new calamity, and it is a beautiful country. The cities of Budapest and Györ, which is our local town, have some breathtaking architecture, wonderful shops for the ladies, street markets, churches, monasteries and stud farms. The latter are relics of the imperial days, when they had to keep the household cavalry supplied and are of a scale and quality that we couldn't believe. Literally acres of deep tiled roofs, providing cool quarters for 200 brood-mares, is not unusual; the horse culture is still very strong and stems from Genghis Khan. The rolling countryside, so sparsely populated, is another world; heavily forested all around the huge arable blocks, there is varied and testing shooting to be had, but usually kept in hand by the owners. In view of the chance of being charged by wild boar, or having a shot at deer or pheasant, the three-barrelled armament (rifle and shotgun) is wise.

The owner of a recently taken-over block is the survivor of an ancient Hungarian land owning family whose holdings totalled 250,000 acres — or was it hectares? (Never mind; that was nothing — the Esterhazys owned over a million.) He works and lives in Munich, only coming down with his friends to shoot occasionally, on the remains of his family estate, which he has managed to buy back.

As I say, a fascinating country.

Chapter 9

Memorable Disasters

In the normal run of farming years one season blends into the next, and small variations in timing or conditions disappear into the mists of time. Even quite major differences tend to even out in the memory, especially if the farmer has not kept a diary — as I did not!

It follows that only major triumphs or disasters mark the progress of time, and in my case the latter seemed to predominate; man-made ones were the main culprits, and of those the ten or so times that we had to call on the fire service come to mind first.

What a customer I was! For my first call I blame some advice given at the Royal Agricultural College, on the treatment and preservation of fencing stakes: we were told quite emphatically that if we stood the bundle of stakes in a barrel of creosote over a fire, the heat would drive the preservative up into the wood, and they would last for ever. Determined to put all my new-found knowledge to work, I set to with a will, and lo, the oil drum split. The creosote seeped and spread across the ground like a volcanic discharge, until it reached the old Nissen huts on either side. These heavily tarred sheds were not slow to respond, and we had a glorious inferno, which the local Sawtry (part-time) brigade had quite a time controlling. It was a pity that they did, as the unsightly pair was the first view that visitors had, and we demolished them anyway. They did say it was a very useful training exercise, but don't do it again!

They consumed a considerable amount of beer, but at least knew where to come for the next call-out. This happened all too soon, when some clever clearing (by match) of broken bales came roaring down the field towards the farm buildings, threatening to demolish everything at the ranch. I must say that the response time of the Sawtry brigade was fantastic once they realised they had a pyromaniac on their patch.

A short pause of a year or three followed, until Mick, the crawler driver, decided that his vehicle with its metal tracks would be immune to simple little heaps of flickering flames as he sought to clear the field of debris likely to block the plough or cultivator. Unfortunately he had reckoned without the plastic fuel pipe, and we had a flaming torch in no time. The Sawtry brigade were there as quickly as ever, but the tractor burnt right out and was a total loss. By now the NFU Mutual were becoming wary of the insurance risk at Grange Farm, and we took great care to avoid playing with matches.

It wasn't always our fires nor our fault. Time and time again roadside stacks of hay or straw would be ignited by passing well-wishers; these posed a tricky problem, as they had to be accessible to the haulage contractors who would be clearing them throughout the winter, and as such were also as easily reached by others less well intentioned. The effect was not always a financial catastrophe, and I know full well that one insurance claim was paid for a stack of best quality hay, when the smouldering heap of grey ash really represented middling straw.

Not everyone was so lucky, and in one horrifying fire at Arthur Burrough's barn at Old Weston, despite the best attempts of all his neighbours, the piglets kept running back into the burning building; no matter how many times we carried them out they repeatedly rushed back to Mother, who had been led or driven to safety, and many were roasted as a result. It was all rather distressing, and we ourselves were scorched and blackened to no avail by the end.

Cottage fires also seem to have figured in several dramas; the old farmhouse at Grange Farm had American service tenants, who left the chip pan on the stove when they went out for the evening. The resulting holocaust nearly destroyed the house, but the dear old Sawtry part-timers had virtually got the fire under control, when the professionals turned up. Frustrated at being denied their glory, they attacked the windows with their axes, allowing a rush of air in to fan the embers and start the inferno roaring again. We nearly lost the next-door cottage twice, when the occupants, different in each case, left their firewood drying in the Rayburn cooker!

The potato-riddling gang proved a very effective substitute for the fire brigade when a smouldering bonfire crept insidiously into a large wood heap. A human chain passed buckets of water from the pond and a flooded shed in such a steady and continuous stream that the bonfire stood no chance. All the ladies of the gang, who were regular members of the lifting team, agreed that it was much more fun than doing the potato job.

Driving the old John Deere combine as part of the B-team with my boys, Ben and Wil, on the trailers, I found myself accompanied by a cloud of smoke. Oh dear! I had not cleared the dust from the engine compartment sufficiently well, and all I had in the way of fire-fighting equipment was a bottle of orange squash and the pint of beer in my own bladder. Having exhausted my supplies, I still had traces of glowing embers to worry about, and set about the journey home to base at full speed. As I reached the yard and safety the flames flared up again, but reinforcements and extinguishers were at hand and all was well. Ben and Wil also distinguished themselves when the whole drier switchboard went up in smoke, putting out the fire and saving the barn; we needed a new control panel anyway!

It was a brand-new International combine, which invented its own system for igniting a field of wheat. Complete with integral straw choppers, it had the equipment and know-how to strike sparks from the odd stone that would somehow get into the works. It then threw out tufts of flaming straw to light the standing corn, but miraculously did not catch fire itself, and chucked out all the potential risk to itself. As Peter radioed frantically for help from base, to no effect at all owing to the hill, our good neighbour, Dick Warrener, came to the rescue with his team. He had observed the whole drama from the other side of the hedge, and it certainly provided a change of entertainment.

Not all our disasters were home-made, and nature was responsible for some fairly exciting situations. Floods have always been a problem at Hamerton, especially for those low-lying houses along the village street; elsewhere I have recounted the tale of the floods of July 1968, just after the major work to end all flooding had been triumphantly finished. Quite regularly, on a ten-to-fifteen-year cycle, this happens, and all the inhabitants threaten to move. The almost universal land drainage that has been put in has resulted in much faster run-off from the fields, and I can't imagine the old village being built now where it originally stood.

The effect on the hill where the farmhouse stands, and also where I now live (nearly the highest point around) is surprising, in that I have been caught out with my house cellar completely full of water, practically to the level of the ground floor. This has coincided with a hopeful investment in wine (Chateau Cissac and Warr's 1977 vintage port in the worst case) washing the labels off and necessitating the owner of the investment portfolio drinking the whole lot, when the original plan was to drink some, sell the rest and perpetuate a lifetime of free winebibbing. At Steeple Gidding the cellar had a substantial mixed selection on the racks, and the occasion was all the more exciting as my dear wife stripped off down to her knickers and gumboots, passing the cases up the cellar steps to her brave husband, keeping safely above the water line himself!

The summer of 1922 is still spoken of as the worst for farming/ flood problems around here; the sheaves of corn were all washed off down the Alconbury Brook and ended up at Brampton Mill in the Great Ouse. However that had followed the worst drought in living memory, when all the stock had to be fed the leaves off the trees, and water carts were sent to the Nene at Barnwell or the Ouse at Brampton. Later on, when we farmed at Biddenham, we had several islands, which would always keep their heads above water and on which the stock could survive, provided we got food to them by boat or tractor! It was amazing to us that the floods moved so slowly; we had two days' warning when the river was out at Turvey, less than five miles as the crow flies across country, but at least three times that on the river's course.

Notable drought years, or very early harvests, include my starter year, 1959, and 1976, which found my combine driver, Peter, away on holiday while I cut a fair proportion of the harvest in his absence. That year we finished combining before the end of the first week in August, but the crops were nothing to shout about. On the heavy land there were very few saleable potatoes, but those on the fens, mostly irrigated, made their growers' fortunes and bought whole farms on the strength of it. I had given up the unequal struggle by then anyway, and would certainly not have been one of the millionaires: potato growers on the heavy land were lucky to sell more than a ton an acre.

Drought in Hungary is another matter altogether, and I have talked about that elsewhere. The extremes of a continental climate make us appreciate just how lucky we are in this country, where total crop failure is a rarity, and starvation a dim and distant part of history.

Continental winters are something we have forgotten as well, but in the first twenty years of my time farming at Hamerton, there were prolonged periods of hard frost with all the attendant difficulties. In my first winter at Grange Farm my son, Peter, was born, conveniently just in time for Christmas, and his arrival was greeted by a very early cold spell, which froze the water pipes and drains. The formidable Swiss 'monthly' nurse who came to help was horrified at the prospect of taking a pickaxe and spade out into the garden to dispose of the by-products of the baby. I expect Gertrude Siegenthaler — I never dared to call her Trudi — still remembers the Halfords for the basic deficiencies in their plumbing.

1962/3 was even colder, and the big freeze lasted for weeks, although not as long as that of 1947. Carting fodder out to the ewes was a chilling task, riding on the trailer of hay with a howling northeaster cutting your face in two. We failed to get all the ewes to come to the trough and lost quite a few from pregnancy toxaemia (also commonly known as 'twin lamb disease') as a result. We had a nightmare task of keeping some sort of water supply running, and in fact failed, as the main pipe to the farm froze underground.

At least half a dozen times in the first fifteen years the snowfall and subsequent drifts completely filled the roads from hedge to hedge, and we would have up to three days of complete peace and quiet until the big diggers or snowploughs got through. Being not even on a B-road meant we were very low on the list of priorities, but in those days there was always one farm in each parish which was supplied with a lightweight tractor-mounted plough; in our case Bill Berry and his son, John, were expected to keep the roads clear and they did. Frequently they were up and down the Hamerton lifelines all night and, as far as they could, kept us free of the major drifts. There was always a problem where the roads ran east to west, and the drifting built up so fast that the US Air Force couldn't get to work from Alconbury to Molesworth or vice versa.

I dreaded one aspect of this: the deep valley gutter on the farmhouse roof filled up, the sun melted the ensuing drift during the day, only for it to re-freeze again at night to form a dam. This prevented the next day's thaw from reaching the downpipe, so I had to slither and skate my way up on to the roof as soon as the snow came. The whole operation seemed to be asking for an untimely end to MJH, but without my efforts the water that built up behind the dam could only escape up under the slates and thence down through the ceilings. There it was collected in a string of buckets, but the plaster also had a tendency to come down with it. Luckily we haven't had that sort of winter for twenty years or more now.

The days of digging out the tractors from snowdrifts seem to have gone, and no bad thing when I think of us all getting frosted lungs and feeling very ill as a result. The last time that happened was when Hughie Miles was with us, so that must have been in 1977/78. We did have one terrible weekend over April 21st/22nd 1984, I think, with sleet and howling northeasterly gales, which drove all the ewes and their young lambs into the most exposed corners. In spite of our efforts we lost fifty-eight lambs, and quite a number of ewes as well. The same storm did no good to the rape crop, which was in flower lying flat under the wet snow blanket.

The main loss has been the skating and snowballing, which seemed to be a regular winter sport for the children. When I was young we all had skates (see school in 1947), and the ponds, old brickyard at Hamerton and other local pieces of water gave us a social focus which was a huge joy to a generation of children. The whole area would foregather on the flooded meadow at Earith, or the Oundle gravel pits, and barbecues, etc., with liberal doses of Gluwein, were enough to make the winter something to anticipate eagerly. Of course we had toboggans as well, but were short on meaningful hills, and had to go to Wadenhoe for the nearest excitement: the run down from the church to the river had a spice of risk attached which made up for the shortness of the slope. For a proper hill we would have to make a major expedition to Burrough Hill near Oakham, which has an ancient fort on top of a 700-foot down.

Will we ever see this again? Well if we do I shall be in no state to enjoy it!

Sale day — 25/9/1996 — MJH leaning on the tyre, foreground, left.

The new lake — 1st winter.

In Hamerton Grove.

Grange Farm from the south.

Sophie and Linseed.

Charlie, who worked for me for over twenty-five years.

Charlie Slack and David Boothroyd.

Albert, who laid over three miles of hedge for me.

Peter, still on Grange Farm after more than a quarter of a century.

Son Peter — Financier.

Son William (and the new lake) — Doctor.

Son James — Kiwi Farmer.

Son Ben and partner Steph — Engineer.

Daughter Bridget — Nature Reserve Warden.

Daughter Rosie — Marketing and Promotions for New College, Nottingham.

Chapter 10

Recreational Training

This is really about shooting, but other sporting activities have crept into the text from time to time, and none deserve their own chapter.

My father retained some of his weaponry from the First World War, but only for show, and my first introduction to shooting was his furious attempts to drive the pigeons off his Brussels sprouts with a single-barrelled .410. The results were always the same; shredded and uneatable sprouts, while a few feathers fluttered down from the (temporarily) disturbed pigeons.

At school I became quite accurate with a rifle, and on one target I had three holes in the bull out of a group of five, and no trace of the other two shots. After earnest debate the judge elected to place the two absentees in the same place, but I was not convinced myself, even though I found it hard to accept that I could have missed the target completely.

Finding myself on the farm with my semi-trained gun dog (she of the flamboyant racing dive at Cirencester), I set out to acquire a shotgun, and went through testing, plating and fitting procedures at Gallyon's shooting school with Ernie — not the Fastest Milkman in the West, the well-known local instructor! He was convinced that I had central vision, and needed the gun with a stock to go with it: they had a lovely pair of Evans for sale, and at a very reasonable price. I carried them home in triumph, but got cold feet at the look of the cast-off stocks. Surely I couldn't be that deformed, I thought, after all I had never had much bother in hitting a ball, and so took this beautiful pair of Evans back to the shop.

What a fool I was: moving unhappily from gun to gun, with ever-decreasing success, I finally ended up with a central-vision Churchill 25, for more money than I had originally paid for the matching pair of Evans. In the meantime I had been welcomed by my new landlord,

G

treated like a favoured son, and always asked to join his shooting parties, which were twice-yearly raids from Fota Island, alongside his mates. They were nearly all landed and titled gentry from the bogs of Ireland, with one or two exceptions from the southwest of this country. What they thought of this ignorant and inaccurate fellow gun I can only guess at, but they were much too polite to say anything and were kindness itself.

I did have an extremely good dog though, and all my silly games in the garden at Withington paid off: Samba was as steady as a rock, and even better, she was a thinking dog. No matter how useless her master was, she was worth having on any shoot, and I gradually got invited around the district and joined the old Kimbolton syndicate, run then by Pat Wood.

I had good days and bad, but mostly I was a pretty moderate shot — meaning moderately bad! I had enough moments or days of triumph to enjoy it and carry on, while wondering why some came out at all. In particular one of my fellow guns at Kimbolton seemed to go the entire season without hitting a single bird. Nor did he get any pleasure from his sport, year after year failing to connect, and having to collect and claim his neighbour's birds in order to have something to show for his cartridge pile. I have had this happen to me once or twice, when I have had a successful drive, and have had the majority of my bag claimed and picked by a neighbouring gun who had not even pulled his trigger on them. It has usually been important men, accustomed to success, who have been unable to accept the idea of a bad day!

Kimbolton, with all its elm trees and before the Grafham Water Reservoir, was a brilliant shoot; it had a very friendly crowd, varied scenery including some spectacularly high trees, and pheasants to match. For the first half of the season there were so many birds that double guns and loaders were the order of the day. I had another reminder of my mistake in returning the matching pair of Evans; I took someone to load for me, often Albert Spring, the Hamerton keeper, and my lone gun got red hot. Lunch was taken in a beaten-up old caravan, which got thoroughly steamed up as we chomped and quaffed our way through our picnic. There was always a bottle of 'Cheskey' (cherry whiskey) provided by Humphrey Whitbread, and a nice mature Stilton from Pat Wood. Sadly, I had to give up my gun in the shoot when the demands of Lloyd's losses outstripped my ability to pay. It is a great joy and interest now that a good friend, Jeremy Marshall, has bought the heart of the old estate, and very kindly included me in his guest list, so thirty years on I can compare and remember the past.

The current management comes out very well, in spite of having lost so much mature woodland — a whole beat disappeared under Grafham Water, and many complete drives vanished with the lost elms. Imaginative new planting has helped re-create and improve these, and it is still a very good shoot.

A temporary hiccup only, though, and I still shot at various local farm shoots thanks to kind friends, and later joined the Hamerton one when it went commercial as an adjunct to an insurance business. That aspect tended to be fraught with danger, as continental customers had different ideas of safety to most of us. One occasion saw a French guest drop onto one knee and level his gun at the beaters in the cover, apparently anticipating wild boar, and later on the same man and his neighbouring gun swung through each other at a hare that ran between them. Why nobody suffered severe gunshot wounds I shall never know.

Albert Spring was a well-known figure in the shooting world and wider afield, from his time at Arnhem, where as a sergeant in the Paratroop Regiment he had led his men out through the German lines to safety, having refused to let any of the other NCOs surrender. He had strong views on everything and had been known to send an illustrious gun home for endangering the lives of his beaters, after they had been shot at twice in one morning!

He was a totally dedicated keeper, whose mission in life, as he put it, was to keep the meat eaters from the seed eaters. To that end he was tireless in watching over his territory; up a tree waiting for a marauding fox, or lurking night after night in the shadows in his van as he anticipated poachers. There are not many gamekeepers who have had a book published, and Albert had three by the end of his life. Dogs could do no wrong, and were usually preferable to humans. Our scruffy little terrier, Scrabble, was angelic most of the time, but once a year he felt obliged to kill a bantam or a duck. "They must have been provoking him," said Albert, even when the poor bantam had tucked herself way back into the straw stack to hatch her eggs.

He retired several times, but finally gave up in the late 80s. The commercial syndicate had packed up in 1980, when Dutch elm disease had killed off most of the main wood, and we were left with a hollow core of thirty acres of young newly-planted trees, surrounded by an outer ring of mature survivors — mainly oak and ash. I took over the running of the pooled land of the Hamerton Estate, now (all excepting my 1,100) owned by the tenants. When I say I ran it, I use the term loosely, as I still did what Albert wanted, of course. It wasn't until his final retirement that I split off my bit, which included the main wood and, with Charlie Slack enthusiastically contributing, was able to change drives round, put in new pieces of cover, and generally experiment. Some changes were notably successful, and others not! However, as is said about shooting, "If you don't pull the trigger you don't hit anything".

With the cooperation and support of the new owners, Corpus Christi College, and their financial backing, we planted ten new spinneys and dug a two-acre lake, in places and alignment recommended by Martin Tickler from the Game Conservancy. The result was some spectacularly successful new drives, but also some extremely awkward corners of once sensible arable fields, and the new lake provided wonderful habitat for a variety of wildlife (as well as water-skiing behind the quad bike for my two young boys).

It was amazing how quickly passing birds spotted the new stretch of water, and we had the inevitable coot and moorhen within days; green plovers adopted us again, and a lone green sandpiper spent a week. The second year found a pair of greylag geese raising a brood of ten young, and the whole family hung around for the next season as well and made rather a mess of the pond. The rough tussocky grass that we had sown round the outside was nothing elaborate, just cocksfoot and timothy, but proved to be a wonderful natural home for pheasants and partridges, and we allowed it to remain undisturbed, only walking through it on the final day's cock shoot.

I was more than fortunate in finding that Charlie Slack was such a keen naturalist, and happily spent his time observing and filming or videoing anything interesting in the small bird or animal world. This was usually in his own time, as he was my sprayer man and one of the two key men on the farm. By juggling his time with that of harvest help, we managed to rear and release over 1,000 pheasant poults and usually about 200 partridges each season, while still getting the land ploughed and ready for the next crop. It was numerically a failure with the partridges, as our recovery rate was less than twenty per cent compared to over sixty per cent with the pheasants, and we managed four good days plus a beaters' cock day.

My attempts at conservation headlands were not without reward, but the main gain was some rather fetching signboards, which I commissioned from a young art student friend. Stuck in the more obvious weed infestations, they showed a green border around a wheat-coloured centre, with a bee or butterfly in the picture, and were meant to let people know that it was all done on purpose. Were they convinced? I know not, but I rated the attempt at least on a par with Dick Beeby's effort: when his drill ran out and left embarrassing strips of nothing across the field, up sprang little white boards, on which were printed 'Seed', 'No Seed', etc. It caused a lot of innocent amusement.

The team of beaters had been coming for ever, and continued to do so, although faces changed as old age and deaths took their toll. I also had a wonderful gang of lady pickers-up, who knew every hiding place on the shoot, and added greatly to the fun and success of the day. Robin Wise (head girl and nationally known field-trial judge and dog trainer), Liz Clay and Barbara Ayres were always professional, but brought a touch of glamour and a civilizing influence to the proceedings, as well as a lot of laughs. They were not fooled by optimistic guns claiming hard-hit birds where there were none. They were also very quick to spot opportunities for changing gun positions and even drives from their grandstand viewpoint way back behind the guns.

On a shoot where many of my guns brought their own dogs, the pickers-up were very discreet in not pinching fallen birds that could be picked from the line, and in fact it is one of my pet hates to have a team of voracious Hoovers clearing everything as it falls. There are many commercial shoots where that behaviour is the norm, and you would be better leaving your own dog at home; but half the pleasure of the day has gone. Watching your dog working and enjoying him/ herself is wonderful, and even Bandit, my current little terrier, has had his moments! I can still hear the bellowing voice of Doug Wise, "Halford! Get your bloody dog back!" Bandit, of course.

After I retired from the farm in 1996, I carried on running the shoot for the new owner and myself, with Charlie's help. It was not easy for Charlie to answer the demands of three masters (the owner, the farms/ finance director, and me), and after another three years we packed it up. The covers we planted are growing well now, and new release pens can always be built — if they are not illegal by then!

Apart from that, for many years I had a gun up in Wallington, Northumberland on a beautiful estate of 17,000 acres (yes, SEVENTEEN THOUSAND), that had been given to the National Trust by Hugh Cheswright's grandfather. A staunch socialist, and a member of the first Labour government, his game book was full of the early political heroes, and recorded the attendance of the first Lord Stansgate (alias Wedgwood Benn). Apart from the rather strange mixed bag, not many conventional game birds were shot, but the odd golden plover, grouse and pheasant. There was also a memorial stone in the park wall, to the leader of the British Communist Party, who had died out shooting there, though not at anyone's hand!

When Hugh took over a part of the estate and started to build it up into a shoot, he faced an uphill struggle; no keeping had been done for at least fifty years, and the main livestock was foxes and corbies (crows). The grouse moors were all planted up with spruce, as were big blocks of the lowland. The early years were a lot of fun, and were generally known as long walks with shouting. We marched up hill and down dale, fought our way through the forty-acre wood and Stinky, and at the end of a good day reckoned our bag in tens. It was a great triumph when we had our first fifty-bird day, and Hugh worked fantastically hard to build it up. He managed to do his feeding and keeper's round before setting off for the bread-winning session in Newcastle, and by the time we had been meeting for twenty-five years it was a really good shoot, with exciting birds at every corner and wonderful scenery all around.

Originally half his guns drove up from his previous haunts around Huntingdon, and we stayed in spartan conditions in the old joiner's shop. The remainder were local, or at any rate from the North, and we had a thoroughly good blend, and really enjoyable weekends. Later on Hugh and Jean moved into The Two Queens, which had been a pub, but became a stately home thereafter, and so instead of the frozen wastes of the joiner's shop we were housed in considerable luxury, either with the two queens themselves, or with Patricia (Hugh's mother), who had been born in the big house and still occupied half the first floor. The annual shoot dinner was held in the gallery in Wallington Hall, and was memorable for several things; a) Gilly Baker Creswell's smart smoking jacket and embroidered velvet slippers; b) the delicious meal; c) the annual awarding of the Bumble Bee Trophy, which managed to work its way round to each member in turn 'for the outstanding shot of the season'; d) the speech, toast or recitation (e.g. *Albert and the Lion,* by Anthony Clay) and, of course, e) the spectacular surroundings

It was very sad when the National Trust made it so difficult to run a shoot for fear of upsetting the other visitors: we must not meet with guns in the main courtyard; we must not shoot the home woods – that finally included the whole park. The final straw was in the aftermath of a big storm, which brought devastation to all our release pens, and nobody would lift a finger to clear fallen trees from our key areas. Hugh gave up in despair, and the syndicate became Happy Wanderers, the "zingari" of the shooting field.

I cannot possibly recite all the memorable days on different shoots that I have enjoyed, but one or two stand out for odd reasons, and several have already crept into the narrative unannounced. Two trips to Brackenbank, Cumbria, a commercially run shooting lodge with (supposedly) two days' grouse shooting each time, I remember well for the almost complete dearth of birds. I shot five cartridges on my first day there, and two of those were at jays. In total over the four days' shooting I saw thousands of acres of beautiful moorland, peregrine and merlin falcons, but fired my gun less than forty times — and not very successfully then! There was good company, excellent food, and it was an all-round experience not to have been missed, though not for the shooting!

A day at Sandringham, previously noted for my disgusting eating habits, where the birds were so wild that they only had to hear a Landrover door shut, or see the first gun approach his peg, and the whole lot would rocket up and home at the speed of light — very testing! I have enjoyed visits to Sandringham under different circumstances, but never have they been so exciting. The livery company visit in the summer of 2001, when Princess Anne was our master, gave a wonderful insight into the workings of the Royal Estate; notable enterprises were everywhere, and ex-crew members of the decommissioned Royal Yacht *Britannia* were very much in evidence. The timberyard was run exactly as one would expect, with every product stacked as if stowed on deck and the ropes all coiled just so. The chief petty officer in charge couldn't be mistaken for anything else, and was delightfully reactionary and royalist in his attitude to everything and everyone. The stud was a model too, and the whole place would have won many farms competitions had it been allowed to enter. From marsh harriers and short-eared owls at one end, to very good arable farming at the other, it was impressive — but my day's shooting there as a guest of Henry Cross is the most abiding memory.

In the ever less populated world of arable farming, the shooting season comes at just the right time: dark days and long nights need something to brighten them up, and the fellowship and fun on a shoot is a lifeline of social contact with humanity. By the time you add up the eight to ten guns, their wives and sweethearts, the pickers-up and dogs, and a cheerful crowd of beaters, neighbouring keepers, cooks, and of course the man in charge (the home keeper), you have a jolly crowd, all set on enjoying their day and each other's company. This applies even if there is not much to shoot at, and there is always a meal of some sort to round off the day's fun. (Ebbe Heyman's pheasants may have strayed over his neighbour's boundary for the shoot day, but the Danish lunch provided by Irene was always worth travelling the length of the country to enjoy.)

Others:

In retirement I have tried to revive my long-defunct golf, and found it to be not so easy the second time around; the swivelling hips, the extended follow through, and the sheer feeling of power at one's command are just not there. Having started with the advantage of being born by the 9th tee of Hendon Golf Club to a mother who played to a 12 handicap, and in her turn was the daughter of a scratch player at St Andrew's, how could I possibly fail to live up to expectations?

Well I did! Learning from a charmingly relaxed old pro at Woldingham on the North Downs, at about eleven years old: Parkes made it all seem so easy, as he put a row of balls down, saying, " Just walk up to the ball, line your feet up, look where you're going to aim, one waggle to loosen up, and hit it." It worked, and the ball went off as directed without much fuss. However, I only got on the course during the school holidays, and reckon I played my best at about the age of twenty-one, when I managed to go round in 9 over par on a good day. I never had the official handicap to go with it though, and more or less dropped the game when I started farming, perhaps playing twice a year. Now I have the time and desire to improve, but old age and decrepitude have a say. What makes it all the more galling is that many of my friends have managed to improve their game to an alarming standard, and I single out my good friend and ex-doctor, Bob Berrington, as chief culprit in this, my humiliation. How dare he!

What about tennis then? Well, yes, I can still hit the ball quite well if I see it soon enough and don't have to run too far to get it. I do regret not having taken my skills more seriously when I was young and sufficiently able at most games to have made something worthwhile to carry into my dotage. Many of the houses at Woldingham had tennis courts, all grass then of course, and tennis parties were almost a way of life. Nothing was taken too seriously, and on the few occasions when I entered seaside tournaments, I met with total humiliation. I still shudder at the thought of meeting the Cambridge University captain in the first round at some pleasant little resort (Budleigh Salterton, I think). I barely got a point, whereas I had been used to just casually waving my racquet, and getting away with it!

Sad to think of the difference between that and the dedication and will to win that had brought my Auntie Phyl so much success and glory in the pre-war years. Phyllis Mudford was my father's first cousin, and playing with great determination and parental support, she got into the last eight at Wimbledon, and actually won the ladies' doubles the year I was born. That was with Mrs Shepherd-Barron, and with a different partner and playing under her married name (King) she reached the finals again in 1937. She also captained the Wightman Cup team, and is our family hero, still going strong in her ninety-ninth year, although tennis came off the menu, as far as her playing the game, when she reached the age of eighty-eight. As for my tennis, I now get a lot of pleasure from helping to run a tournament in aid of Macmillan Cancer Relief, each year. With the use of up to ten courts and with between sixty and eighty entries, it can be a nightmare, as people drop out at the last minute, turn up at the wrong court, take exception to their partner or the scoring system and various unforeseen slips. In the end, though, most have a good time, and Bill Godfrey and myself (with Lucy's help and input) feel that it has been worth it, and some £1,500 has been raised each year for a very good cause.

My first conscious effort with a cricket ball was at my junior school sports, where I flung it — in the competition mind you — miles into the crowd. It may have been the longest throw by a street, but was so off line that it didn't count, and in fact very nearly brained an innocent bystander. Only the quick reflexes and catching ability of my own father prevented an embarrassing accident; but never mind, I won the eighty yards sprint, and flushed with triumph, managed to crash my bike and knocked myself unconscious as I hurtled back down to the prize giving.

At the age of thirteen, and with a deadly slow off-break, it was probably not the most tactful career move to hit my headmaster's middle stump first ball. However, it was about the pinnacle of my success, as my slow off-breaks were murdered in the tougher competition of older boys at Winchester. Attempts to speed up my delivery resulted in losing both flight and turn and I became even easier meat. With a good eye, I should have made something of my batting, but never managed to achieve much in that line either. Oh well, it's too late to put that right now, and the it-might-have-been syndrome is unproductive and depressing.

Even singing is something I have left too late to make much of. I was blessed with a good treble voice, and sang in the chapel choir as a bass. When I worked in London I managed to pass the audition for the Bach Choir, probably because at the age of twenty-three I was not past all hope of improving. I sang with them for several years, which involved Monday evening practices at the Roman Catholic Cathedral Hall, and a late train home to bed at Caterham. The concerts at the Albert and Festival Halls were inspiring events, and I got so carried away in one carol concert, that I misinterpreted Reggie Jacques' final flourish to *Masters in This Hall*, as an extra (third) triumphant chorus. I bellowed out "NO" as my preliminary to *Noel, Noel, Noel Sing We loud!* only to hear my solo effort echoing around, luckily still swamped by the orchestral reverberations.

Leaving the city life for farming made the weekly practice very difficult, and from the Royal Agricultural College it was impossible. I did try again when I got to the farm at Hamerton, but catching a train up to London at about 4.30 and arriving home at midnight proved too much for me and my new farming life, and for nearly fifteen years I forgot about singing. A chance encounter at Great Gidding Church, where Mike Keck was recruiting for a local concert, started me off again, and led me into the Huntingdon Choral Society (now the 'Hunts Phil') where I have enjoyed the music and social life enormously.

I blame Ken Peggs for most of this: as a keen young bank manager (not mine!) newly appointed to Barclays at Huntingdon, he would try anything for business and turned up at the Hamerton village fete, in support of his wife who was the official opener. Janet Peggs may have entered the shapely ankles competition, but it was Ken, with his third place in the knobbly knees, who attracted the most attention. Stung at being beaten by me and my Scots farming neighbour, he got his own back by pressing me into joining the choral society. In those early days we gave concerts in freezing churches, some so cold that we had to carry hot-water bottles and hip flasks, and our audience brought rugs and blankets as well.

At Ramsey Church the afternoon practice before one winter concert was so paralysing that many of us managed to squeeze extra sweaters under our evening shirts for the concert. To arrive at 7.30 p.m. and be greeted by the blast of a jet-engine heater, and a temperature approaching the 80s, gave us no chance to modify our attire, and we threatened to expire en masse. I also have the strange recollection that at no moment during the entire concert did I find myself singing the same note as my neighbouring bass. The warmest church, the Free Church at St Ives, became our regular concert venue for quite a time, but there we had to assemble and dismantle the staging for every concert, which tended to prolong the evening tiresomely, especially for any 'fans' we had dragged along. The staging also provided excitement and a spice of danger for the back row of the basses as we tried not to fall back through the windows onto the pavement behind and below.

Thank goodness we now have The Mary Stuart Hall at Hinchingbrooke, alias the Huntingdon Performing Arts Centre with comfortable tiered seating for both choir and audience. What wimps we have become!

With some coaching and a bit of effort I should have been able to sing solos, but when it came to the crunch I was strangled by nerves and produced the most awful sounds. So I have remained in the back row of the basses, and get a lot of pleasure from mooing in the herd! Currently we are getting ready to be recorded for the Advent Sunday *Songs of Praise* broadcast from Godmanchester Church. (I write this on October 23rd 2003.) This was the scene of some of our earlier frostbitten sessions, and also where we sang for *Songs of Praise* twenty-seven years earlier. The voice is weaker, the range has foreshortened, and I puff and blow as I struggle to keep up with the flying pickets. The fast runs favoured by Handel are a particular hazard, and one day I shall be caught out for singing only the first note of each run of four, or some other heinous offence. Oh well, it has been good while it

Bandit (his master's odour-free feet).

Chapter 11

Back to the Pen

I have found myself compelled to launch into print from time to time, in response to attacks in the press on farmers in general, and certain practices in particular. Also as chairman of the Tenant Farmers' Association, I was forever sticking my head above the parapet.

I have included a small appendix section with some examples, which remind me of battles fought in the past, and may jog others' memories. However, that is all history, and I now find myself becoming ever more fascinated by the local records for the parishes of Hamerton and the different Giddings. Having researched the decline and virtual disappearance of the village of Steeple Gidding, I produced a booklet which covers the last hundred years, and this has so far sold over 300 copies, mainly to people visiting the church and those with family ties in the past. Now that I have all the census information available, and copies of the parish records of baptisms, marriages and burials, I am better able to answer the enquiries that arrive with surprising frequency.

For instance, a Mr Gilby wrote in, and I was able to flood him with details about Gilb(e)ys past; one poor couple have six of their young children buried in the same grave, and a common factor is the very high death rate in young children.

In the past, the estate was owned by the Cotton family, major landowners and powers in the land in the 15-1600s. There was a stately home in the grass field opposite the church, but that fell into disrepair through neglect, when no member of the family chose to live here (having quite a choice of big houses in the area). Conington Castle, now demolished, was their main seat, and a bequest of the antiquarian books and papers collection formed the foundation of the British Museum.

There is a magnificent wall tablet, detailing the family link to Robert the Bruce, which I have included in the appendix; it goes into such detail that the thought crossed my mind 'Who are you kidding?'

From past history to outright tabloid journalism is but a small step, and so I now find myself writing up local functions for the parish magazine; nobody is safe from your correspondent, and the writs are building up fast. Even my own forebears are dragged into the limelight, although for family interest only.

The black sheep that crop up in every branch and each generation prove far more entertaining than the rather fewer saintly members, and I look forward to many happy hours exploring the past. I have already attempted to write up the family history for the benefit of my children, and tracked all four lines back from my grandparents as far as I had time for then. There is already something about the Monros, my mother's family, in the 'Family Business', but currently exercising my mind is the question mark over the rather dubious clause in my great-grandfather, General Harrison's will, in which he refers to his 'reputed children by my second wife'. As one of these reputed children was my own grandmother, Ethel Monro, I would like to know what he meant! The great-grandmother in question was one of two gorgeous Brown sisters, who set hearts aflame at the time. Daisy became Lady Bonsor and one of King Edward VII's favourites; Eliza, Mrs Edward Harrison, may have strayed, and certainly disappeared without trace as far as any records go. There was a very fine collection of French Empire furniture, with beautiful marquetry and enamel panels, which was always passed down from mother to daughter, and this came from where originally? I do not know, nor did my mother while she housed it, nor my sister in her turn. What admirer gave that outstanding present? What happened to Eliza? and interestingly, how come that Daisy sent a floral tribute with her love, on the general's death in 1909, if her sister had been kicked out?

John Fenn Halford, my great-great-grandfather, an apparent pillar of society and master of the Worshipful Company of Ironmongers (although a stockbroker in fact), only married the mother of his four children, Frances Amelia Brooks, when the eldest of them was twenty-four years old, in 1845; and shot up to Gretna Green to do it. We have a copy of the marriage certificate from Dumfries Record Office, but also have a copy of the licence taken out for their wedding at St Michael's, Cornhill the day before. What is going on? Is there another Halford family lurking in the woodwork? For a while I thought I had the answer, when I found the will of Edward Brooks, whom I took to be the first husband of Frances Amelia (JFH's wife and my great-great-grandmother). The will leaving his wife, Frances, substantial property and income only if she did not re-marry, seemed to provide a strong reason for the late marriage (i.e. after the executors and alternative beneficiaries had died). Sadly this turned out to be a red herring when the baptism of Frances Amelia Brooks was found where it should be: if it was the widow she would have been christened under the name of Fryer.

The Halford line had been a dead end for many years, as far as I knew; I always believed that I was the only son of an only son of an only son, etc. No mention was ever made to my grandfather, referred to as 'The Cad and Bounder', after I learnt that he had only married my grandmother, Julia Isabella Anstey, four months before she gave birth to my father. Having been pushed into a shotgun wedding, he then bunked off never to be seen again. However, I did spot a notice from the Treasury solicitor in 1958, asking for any relatives of John Oswald Halford (deceased at Yarmouth) to get in touch. As my father was John Alfred, I thought this attraction to Anglo-Saxon names was likely to be interesting, and so it proved.

To cut a very long saga short, we found that John Oswald had been one of a family of eight, and that his parents had emigrated to Australia after their marriage in 1852. There was Alfred Brooks Halford and his wife on the passenger list of the *Strathfieldsaye*, bound for the Victoria goldfields, landing at Port Melbourne in April 1853. Sadly he did not make his fortune, and came home ten years later as his father was dying. His father was John Fenn — he of the mysterious Gretna Green marriage!

Set alight by all this excitement, I now pursue every Halford looking for links. For a start there was the schoolboy who carved his name into the old cloister arch at Winchester in 1668! (Vandals to the core, we Halfords.) The school records show that two brothers were at the college, as 'founder's kin'; i.e. they were direct descendants of William of Wykeham, and came from Welham and Wistow, near Market Harborough, and the family were staunch royalists and created baronets for that. All very fine, but so what if I can't tie up any connection? I haven't given up yet, though, and Blue Mantle pursuivant may yet hear from me!

Then there is a plethora of Halfords in the Melbourne telephone directory; did some of the brothers, cousins or whatever return to Australia, and make fortunes? Possibly not, but still . . . . wishful thinking.

F. M. Halford, the great authority on dry-fly fishing, turns out not to have been a Halford at all, but changed his and all his family's name from his original German for business or social reasons. His grandson turned up at my prep school in my last year at Sandroyd, and has sadly just died.

The Anstey line also provided plenty of amusement: Julia Isabella originated from her grandmother, a Portugese lady who married the postmaster general of Jamaica. Legend has it that she was very skilled at origami, and that nothing was safe from her lightning scissor cuts, including the title deeds to the estate/plantation in Jamaica.

The postmaster Anstey was the grandson of Sir Christopher, society scribe and poet, who chronicled Regency Bath in rather dreadful verse. He made a lot of money in doing so, and is prominently remembered in Poet's Corner, Westminster Abbey. He had chosen Bath rather than remain at Anstey Hall, Trumpington, which fine property had been acquired by extremely questionable means, by his father — a reverend as well.

So is anyone's conscience clear, or are we all blackguards, and it is all the fault of our genes?

Now I come to this little effort, which is a random selection of snapshots from my life so far; it is not a continuous narrative, which the readers who have persevered thus far will have deduced for themselves. Many of my closest friends have been left out in the cold, and probably thank me for that. I had many completely satisfactory and happy business relationships throughout my time farming, and the people whom I dealt with over the thirty-seven years are still good friends. My six charming and thoroughly delightful children are only mentioned in passing because I don't want to be killed by them, and proud parents can be very boring.

I hope you, dear reader, have not been too bored as you ploughed through, but you could always have put the book down!

Appendix

Letters from myself in defence of . . . .

Shearing Sheep (9th June 1972), written in answer to a Mrs Hodges of Oundle, who complained of the condition of a flock after shearing: 'Every sheep was limping or bleeding from cuts'.

I was distressed to read that one of your readers considers us cruel and inhumane when we shear our sheep. I would like to reassure her a little, if possible, and perhaps convince her that we do not intentionally take strips of sheep's hide for bootlaces.

To understand the shearer's problem one should start with the intellect and psychology of the sheep itself. Domesticated as it has been for thousands of years, you might suppose that the animal would regard man as a friend, and that an understanding would have developed; alas, little sign of this bond is evident.

Spend infinite time and trouble growing an expensive and delicious field of mixed grasses and clovers, and the sheep spends all her waking hours in trying to force, jump or crawl her way out in order to dance on your prize crop of winter wheat next door. Smile winningly at her, inviting her to take her lovely medicine, and she will leap your head to avoid her vital worm drench. Try to trim an overgrown hoof to cure her crippling foot rot, and she is miraculously better and able to run like a stag.

It follows that shearing is an even more traumatic experience for the poor animal; she need not take it lying down, though, and for a start often weighs as much as her shearer, so can quite easily get him off balance and fling him to the ground, where he desperately tries to avoid the wildly gyrating clippers. By squirming cleverly in a sort of porpoise roll, she can find herself free to rush between the poor man's legs, carrying him off on her back and trailing yards of half-shorn fleece behind. The clippers meanwhile are free to perform their dance of death on the floor.

At the end of the day, the shearer is bent double, he believes forever; he has several fingers plastered up; he has been kicked painfully where he would not have chosen, and to cap it all he finds he has run out of beer. For most of us this is the normal pattern of shearing time; it is the job, out of all those on the farm, that requires the most skill, stamina and sheer hard work.

There are gifted individuals, mainly from down under, who only have to wave a magic wand and lo, 400 sheep walk out at the other end, painlessly and perfectly shorn. They write books about it, full of helpful diagrams, but for most of us the job remains as it always has been; sheer hard graft. So next time the flock goes past your correspondent could she please spare a thought for the crippled and bleeding individual at the back.

The straw burning controversy; there were pages of letters complaining about this practice. This reply was printed in the Farmers Weekly, surrounded by the opposite viewpoint.

As a farmer with over 1,000 acres of straw crops in Beds and Hunts, I have contributed my share of smoke and smuts to your summer. That is not to say that I can honestly defend the practice of straw burning, and certainly that of hedge and tree scorching that also takes place as an unwelcome side effect. It saddens and sickens me to see what some of my fellow farmers do to their surviving hedges and trees, having still about twenty miles of hedgerow here, which I have defended more or less successfully for over twenty years.

However carefully I burn my surplus straw, though, weather conditions alone decide if the smoke and smuts are going to mess up your clean paintwork. To Canon Roy Meadows and all your other correspondents, I apologise; we all of us know in our hearts that we are being anti-social, but can we look at the reasons for it and the alternatives.

This area is one of the main breadbaskets of England, and in Huntingdonshire alone there are some 40,000 acres of cereal crops, representing 80,000 tons of straw at least. Nearly 5,000,000 bales, of which less than 10% is needed for the livestock in the region; about the same again is taken west (at great haulage cost) or to processing plants at Kimbolton and Warboys. So that leaves approx 4,000,000 bales which nobody wants.

Canon Meadows asks about the possibilities of using it for fuel, and indeed this is a valid point. Unfortunately the capital and running costs of the cubing/briquetteing machinery makes the resulting product far too expensive to compete with wood at one end of the scale, or Phurnacite at the other.

The last alternative, that of incorporating it into the soil, is the subject of intensive research going on at various experimental husbandry farms, and on light, easily worked loams the procedure can almost be said to be easy. For those of us farming the heavy boulder clays of West Cambs, Beds, Hunts and Northants, it is virtually impossible; if it can be done at all, it requires endless passes with machinery; it costs

H

a fortune, both in work and in lost yield. This particular season with its drought conditions, we have had to struggle with a tilth consisting mainly of baked clay lumps from 6" across, upwards. The wearing parts on our cultivating machinery glowed red as they worked, and replacements became virtually unobtainable.

Work is going on in the machinery and chemical worlds to try to find solutions, and the problem may be solved soon. As I said, I am not writing to defend the practice, but to pinpoint some of the problems, and apologise; please accept these, Canon Meadows and all you good people, and try not to hate us too much.

READERS' LETTERS

World prices are threat to small farms

Sir, A broadside from Vincent Hedley-Lewis (Letters, Feb 21) is enough to leave most ships sinking, but before I jump into my dinghy and paddle off, may I be allowed a last few shots from my two-panels on the foredeck?

Much of what Mr Hedley-Lewis says is hard to refute, and it comes down to: What have we to be afraid of in telling the truth that we can and do produce at world prices. Who can possibly fear that?

Well, for a start, the smaller acreage man who just happens to be the average farmer — a one-man-band with some family help, producing 500t of wheat, or the equivalent in other crops. If he takes out quite modest living expenses and tax of £1500 he is effectively adding £33/t to his growing costs.

Another who needs to worry is the tenant, as we both agree. His wheat will cost up to £15/t extra and more than 30% of us are tenants. What about our landlords? They also need an income, whether private, individual or pension fund. They will have to get used to the idea of lower rents, but these will not come down overnight, and nor will they reach the zero figure that would be needed to put us on an equal footing with the owner-occupier.

The world price claim is dangerous to us all, because it leaves no room for less productive or failed crops, waste land, and certainly not 15% compulsory set-aside without compensation. I, too, had a remarkable margin for my linseed, but without the acreage payment it would have looked a sick joke. Rape

this world price argument leads u
The total acreage payments on m
farm will average over £100/acre;
how long will that be politically
sustainable if it is perceived as be
over and above the cost of
production?

The environment will be another loser; the decline of the small farm and small workforce will be hastened. The 100-acre field, as opposed to 4 x 25, already receives more acreage subsidy as the crop

8

itself is measured — no more room for hedges and rough ground here. I, too, am an admirer of Oliver Walston and Chris Hollingworth, the latter as a fellow-member of the same agronomy group. I fully realis the flaws in the present system, but I do see dangers for a large section o the farming community if his figure are taken as the norm.
Michael Halford

Stop over egging the custard, Jeff

Anyone reading Jeff Swift's article (FG, August 7) would go away with some very clear beliefs, not all based on fact.

These include:

Jeff Swift says everything was fine before the change in the Agricultural Tenancy Law relating to succession — that the landlord/tenant system still worked, so why change it?

Those of us fortunate enough to be renting farms in 1976 were presented with a further two generations, and there are some very good young farmers where they are today because of it.

There are many times more who have been unable to get a farm. And hundreds of thousands of acres that have not been let. We cannot avoid the fact that, in the real world, those many landowners who want to let are doing so — but through loopholes in the law, contrivances like share farming, as well as Gladstone-Bowers and Ministry licences.

Handicaps

He says that owners can be compelled to let their land, and what's more, on terms of the tenant's choosing. Come, come, Jeff — of course they can't, and that is why we have sought to eliminate the more obvious reasons for not doing so. Succession was a major one and, (such as a new business).

We must not lose sight of the main objective; to bring land on to the letting market, not only for new entrants but for the sake of existing farmers as well. County council tenancies are important, but many county council tenants want to expand, or move on. The law must serve the purposes of both sides.

Joking

He suggests that the farming organisations at best cave in under pressure, and at worst do nothing. Oh dear. When I think of the hours spent, the MPs lobbied, the miles travelled . . ., but he must be joking. We did have a very fruitful meeting with the Minister only two weeks ago, at which Tenancy Reform was the top priority. Productivity and related earning capacity was one of our successes too.

We hope that the Ministry's proposals will be a great improvement on the consultation paper, because we have met them with fact and argument about today's needs, rather than simple resistance to change.

So, Jeff Swift, while many of your aims are sound — and even some of your ideas for achieving them — could we have a little less egg in the custard, please?

Michael Halford,
Chairman, Tenant
Farmers' Association,
Grange Farm,
Hamerton,
Huntingdon,
Cambridgeshire.

Tenant
Farmers'
Association

7 Brewery Court, Theale,
Reading, Berks. RG7 5AJ.
Telephone: (0734) 306130
Fax: (0734) 303424

Mrs. Gillian Shephard M.P.   27th. May 1997
H.M. Minister of Agriculture
Whitehall Place
London W.1

Dear Mrs Shephard,

I should like to welcome you to your new post at Agriculture and
congratulate you on your appointment, on behalf of the Tenant
Farmers Association and myself.

We enjoyed a friendly and helpful relationship with your
predecessor, which yielded many benefits and promised more to
come. In particular we were a long way along the road towards
Tenancy Reform, with the vital and overriding need to halt the
continuing decline in the Landlord/Tenant system and once again
open up the market to young farmers and new entrants.

We had achieved virtual agreement within the industry: the C.L.A,
N.F.U., Y.F.C, and ourselves are all prepared to accept the
necessary safeguard clauses,and only one point-that of the need
for a minimum term required by the N.F.U.-stood between us and
unanimity.

May I urge you to pick up this ball and run with it, if that is a
suitable metaphor, as soon as you have had a chance to settle in.
We at the T.F.A. shall be as helpful and constructive as we
possibly can, and I look forward to meeting you soon and working
harmoniously with you in the future.

Yours sincerely

            Michael Halford,Chairman

MAFF

Ministry of Agriculture, Fisheries and Food
Whitehall Place, London SW1A 2HH

From the Minister

M Halford Esq
Chairman
Tenant Farmers' Association
7 Brewery Court
Theale
Reading
Berks
RG7 5AJ

11 June 1993

Dear Mr. Halford

Thank you for your letter of 27 May and the kind words of welcome on my appointment.

You particularly stressed the need for new agricultural tenancy legislation. Coming from a farming background and a farming constituency, I am certainly conscious that there has been a long-term decline in the tenanted sector which nowadays makes it more difficult for young people to get started in the industry. This decline can only be reversed by making it more attractive for landowners to let land, which of course is why reform is needed. I was therefore encouraged to hear that there is a good measure of agreement between yourselves, the CLA, NFU and National Federation of Young Farmers' Clubs.

However, the introduction of new tenancy legislation is, as always, dependent on Parliamentary time being available and, as you have no doubt heard, there is very great pressure on the Government's legislative programme for next Session. I hope to make a statement on this subject fairly shortly.

I would be delighted to meet you although, as I am sure you will appreciate, my diary is somewhat congested at the moment. I will ask my diary secretary to contact you in the next few weeks to see what can be arranged.

Yours truly

Gillian Shephard

GILLIAN SHEPHARD

Ministry of Agriculture, Fisheries and Food
Whitehall Place, London SW1A 2HH

From the Minister

The Rt Hon John Major MP
Prime Minister
10 Downing Street
London
SW1A 2AA

Our ref: 74892

27 January 1994

Dear Prime Minister

Thank you for letting me see the letter of 9 December 1993 from the Chairman of the Tenant Farmers' Association, Mr M Halford of Grange Farm, Hamerton, Huntingdon about the reform of agricultural tenancy law.

I am delighted that it has been possible to find a common position which commands the clear support of representatives of all interests in the agricultural industry. I am in no doubt that the agreement is in the best interests of both landlords and tenants alike and the wider rural economy will benefit from these far-reaching proposals for reform.

The fact that such significant progress has been made within the agricultural industry is due in no small measure to the efforts of Mr Halford and his colleagues in the Tenant Farmers' Association. The TFA have been a positive influence throughout, helping to ensure that discussions have been conducted in a constructive manner, both with other industry organisations and with my officials. I am grateful to the TFA for all that they have done.

I am anxious to introduce legislation as soon as Parliamentary time becomes available and I shall be making all the necessary preparations to that end. As Mr Halford says, this is an excellent opportunity and we must make sure we deliver the reform that is so badly needed.

Yours ever
Gillian.

GILLIAN SHEPHARD

From: The Rt Hon John Major MP

HOUSE OF COMMONS
LONDON SW1A 0AA   31 January 1994

Dear Michael,

    Thank you very much for your letter of 9 December which was acknowledged by my secretary and which I have read with care. I am sorry for the delay in replying to you myself.

    I have now heard from Gillian Shephard at MAFF, to whom I sent your letter for her comments, and I am enclosing her reply to me which is self-explanatory.

    I hope you find the Minister's comments both helpful and reassuring. I know that she does very much appreciate the assistance which you and the TFA have been able to give during the course of discussions on this very important matter.

With best wishes

Yours sincerely,

John M.

Michael Halford Esq
Grange Farm
Hamerton
Huntingdon
Cambs PE17 5QW

EAST OF ENGLAND AGRICULTURAL SOCIETY

FARMS COMPETITION 1993

Generously sponsored by
Sidney C. Banks p.l.c.
Grain Merchants and Seed Specialists
of
Sandy, Bedfordshire

JUDGES' COMMENTS

CLASS B   NO OF FARMS 6

JUDGES   Mrs Sarah Ward
           J Scott Esq.

COMMENTS ON THE FARM OF:

> M J Halford
> Broadroad Ltd
> Grange Farm, Hamerton, Huntingdon
> Cambridgeshire PE17 5QW
> Tel: 083 23 223

Grange Farm was an object lesson in keeping a healthy, interesting farm going with a minimal labour force. The crops looked well — in particular the wheat and linseed — and the policy of no second wheats (tempting on this good ground) had paid off.

Beans looked good except for a few compaction problems on the headlands.

Weeds spoilt some potentially high-yielding rape — no doubt it will combine out very satisfactorily — but could cause drying difficulties.

The ewes and lambs looked in super condition and the management system, though simple to suit the labour provision, was obviously more than adequate. Good choice of breeding ewe and terminal sire. Sold at optimum weights.

The range of flora throughout the farm was impressive — showing how high yielding cereals are not necessarily incompatible with conservation provided the manager knows what he is doing and thoroughly understands what he is aiming at. One did not feel the environmental considerations on Grange Farm were merely cosmetic — they were part and parcel of the whole approach. The grass rides were a pleasure to drive along — bird, animal and insect life appeared as if they were keen to partcipate in the competition.

A farm where as much was given as was taken out. Could he have made more out of the shoot given the potential on the farm.

It was a pleasure to see a farm of high arable potential put to good use. Standard of crop management was high and output no doubt high also. The policy of growing all first wheats had clearly left the land in good heart. Break crops were grown skilfully and the beans and linseed looked particularly well. Some weed problems in the oilseed rape will probably not affect final yield though I was interested to note that the farm relies on a cyclone cleaner. A clean sample off the combine is the mainstay of success here.

The sheep flock looked very well indeed and the excellent lambing percentage is evident of a high standard of stockmanship.

A careful programme of hedge maintainance and a continuing small scale annual tree planting, as practised on this farm, is an effective way of permanently enhancing conservation on this farm. Choice of site for the new plantations was not always obvious but may have been dictated by the way game is expected to fly.

The size of the enterprise and small labour force point to very effective use of a small staff.

Macmillan

cancer relief

a Voice for life

Macmillan Cancer Relief
32 St Andrews Street
Cambridge
CB23AR

INVITATION TENNIS TOURNAMENT

in aid of

MACMILLAN CANCER RELIEF

On Thursday August 16th 2001 at **Hemingford Grey**.

Timetable: Assemble at Hemingford Grey Tennis Club @ 10 am to draw for partners, in groups of up to five pairs, which will then play all against all within each group. Matches will be of seven games and each game counted. We then adjourn for lunch at Bill and Lucy Godfrey's house (The Glebe, 30 High Street), by kind permission, with the picnic that you will have prepared for yourselves (soft drinks will be provided) for a period of R and R and social chit-chat. There is also a swimming pool that those who are not worried about the effect on their game are welcome to enjoy.

The winners of each group, plus the top-scoring second if needed to even up the numbers, will then be despatched back to the club to fight out the semi-finals, and the resulting finalists will come back to the Godfreys' for a grandstand finish on the Centre Court.

The club has three courts, and so the initial capacity for this occasion will be for 40 players. If there is a strong demand for more, we will attempt to accommodate the surplus by finding additional courts, but at the time of going to press that cannot be guaranteed, and acceptances will be on a first come, first served basis.

The entrance fee will be £10 per head, and the prize money negligible, although it is hoped that some token prizes will be donated. It is also hoped to have a raffle to raise more for the fund.

This is intended to be a fun day out for all, with the tennis not too serious, and not too bad either. It should raise several hundred pounds for a very worthy cause. The money will go to providing startup funds to launch two additional Macmillan nurses based at Hinchingbrooke Hospital.

Please send your entry with £10 to: Michael Halford, The Old Rectory, Steeple Gidding, HUNTINGDON, PE28 5RG. (Tel: 01832 293488)
. . . .

I would like to enter the Hemingford Grey Tennis Tournament in aid of Macmillan Cancer Relief, on August 16th at 10 am, and enclose my cheque for £10 (payable to Macmillan Cancer Relief).

I regret I am unable to enter the Tennis Tournament, but enclose a donation/will donate a prize.

NAME..

Macmillan
cancer relief
a Voice for life

Macmillan Cancer Relief
32 St Andrews Street
Cambridge
CB23AR

HUNTINGDONSHIRE INVITATION TENNIS TOURNAMENT

Dear

This year we hope to widen the scope and appeal of the tournament, and have the use of extra courts to allow over 80 entries. In addition to last year's quota we have been offered the two club courts at Papworth and another private court at Hilton, hence the new name above. There will also be more matches in the afternoon for those who do not get through to the second stage.

To try to capture the agricultural vote, we have aimed for a narrow window of opportunity between the rape and wheat harvests. To avoid the grouse season, the Game Fair, Wimbledon, family holidays, Goodwood week and 101 other good excuses is almost impossible but we have booked **Thursday the 25th July**. For those of you who did not play last year, here is a brief outline of the *modus operandi*:

9.30 a.m. Assemble at the two clubs, (you will be told which in advance) to be allocated partners. You will not be playing with your spouse nor regular partner if we can avoid it, and the management wish to make it clear that only an ENORMOUS bribe will affect our judgement (bearing in mind the limited supply of Greek gods and Kournikovas)! There will be up to 5 pairs on each court, playing all against all in matches of 7 games, with each game scored. After this exhausting morning we all repair to Bill and Lucy Godfrey's house, @ 30 High Street, Hemingford Grey for R & R, picnic, swim, *etc*.

The winners of each group then return to the fray to produce the champions, whilst lesser mortals have a go at each other in a 'plate' competition to be devised. Prize money is negligible, (nay nonexistent), but the honour and glory are huge. Hopefully a good time is had by all, and a decent sum raised for a very worthy cause. With that end in mind we thought that participants would not mind an increased entrance fee of £12 per head.

This advance notice should give you all plenty of time to plan your entire summer programme around this event, so please put it into your diaries now, and register your interest by returning the bottom section of this letter to me, Michael Halford, The Old Rectory, Steeple Gidding, Huntingdon, PE28 5RG (01832 293488).

. . . .

a) I would like to enter the Huntingdonshire Macmillan Tennis Tournament on July 25th, & enclose my entry money (cheque payable to Macmillan Cancer Relief)
b) I hope to play and will confirm soon
c) I regret that I definitely will not be able to play
d) I enclose a donation or will give a prize

NAME and contact telephone number . . . .

Please advise any inaccuracies on the address label

HAMERTON MILLENNIUM RECORD

A thousand years ago an inventory of the whole country was ordered by William the Conqueror, and the result was the Domesday Book. The time span as we seek to do the same today is almost past our comprehension, but when one considers that on average there have been four generations in each century, that adds up to forty lives. Eudo of Normandy, granted the estate after 1066, would have been just as daunted at the backwards view of the previous thousand years, and with far less written evidence.

A huge canopy of green covered the country originally, and the great midland forest of Bruneswald (the Saxon name for it) stretched from Buckingham to Bedford, across to join the Herts/Essex weald and up to skirt the fens to Peterborough. The clearance of settlements was a slow and laborious business, helped by rooting pigs and the Saxons' ability with the battleaxe, and several hundred years passed before the area was anything but an almost impenetrable barrier to travel. Evidence of the woodland past can still be found in the place names of Leighton Bromswold, (W)old Weston and others.

Roman Britain was naturally settled along the river valleys and easier working land first, and there are plenty of traces of their occupation, apart from the road network. For instance, when the gas pipeline went through Grange Farm, sites were located by MRI scanning, and researched. Saxon farmsteads were identified and a lot of Romano-British pottery of about AD 200 found. In fact the Nene/Ouse valleys were then settled at almost the same level as in 1900.

Eudo, known as Dapifer (the sewer) to his friends, was probably not so called because of the main drainage he installed in Hamerton. One can only imagine his state of personal hygiene, to arouse comments in an age when bathing was virtually unknown. Over 400 years later it was said of Queen Elizabeth that she took a bath every month whether she needed one or no, and our Eudo was obviously notable in one respect anyway! His predecessor, Ulfkell, was a supporter of King Harold, whose Danish forbears had taken the land by force 200 years before that, from Aelfrith (let us call the Saxon). He in his time had taken it from Trog the Brit, and so on.

It all seems so tame now, but rape and pillage were a regular hazard, and we were very near to the heartland of the Iceni tribe and Queen Boudicca. Our predecessors undoubtedly went to war with her against the Romans. The culture of fighting and dying for your lord and master was the way of things, and it makes the handing over of an arm and a leg, by way of rent to a commercial landlord in more recent times, quite painless in comparison.

In the expectation that his reward would be in heaven, Eudo gave the right to the income from the estate to Colchester Abbey, which owned it until the dissolution of the monasteries by Henry VIII. At that time the pattern of farming was large areas of sheep walk, with small pieces of communal strip farming for arable crops on the kinder land. The site of the original Grove Farm was then on the Winwick side of the valley, where traces can be found now, but most of the other farmsteads were in the village of Hamerton itself.

When the then owner, in 1638, decided to enclose the land and parcel it out more logically, new houses were built at Grange Farm (then Lodge Farm), Salome Wood and a new site at The Grove. All three were built to the same design and out of the same red brick, and along with them were erected new barns and farm buildings: the old hay barn and cart hovel at Grange Farm date from this development, but a similar barn at Grove Farm blew down, sadly, in a freak storm in the 1970s. On one of the timbers was chiselled the date 1638, but there was too much broken to salvage the building as a whole. (Incidentally, quite a number of the timbers were used in the reconstruction of Poet Rowe's House, the home of John and Barbara Ayres in Glatton. That house originally stood in the Bedford area, and was taken apart and moved when threatened by a road development.)

Rookery Farm, known as Town Farm on the early map, was a pre-enclosure site; so was Church Farm where the present rectory stands. Brickyard Farm (now called Church Farm) seems to have arrived when the brickyard itself was opened up for the wave of building in Victorian times; this accounted for the new and grander Grange Farm house, Cottage Farm and many new cottages at Grange, Salome Wood and Manor Farms, as well as the new rectory and school. Incidentally, the owner of the brickyard was a Mr Wadsworth, whose family still runs the wine merchants in St Ives. The suspicion must cross one's mind that drink has always been more profitable than building or farming!

The pattern of tenant farmers and a fatherly but absentee landlord carried on for 300 years, and there were eight separate farms as recently as 1950. The Knightons and Jellis were well-known and respected farming families in the first half of this century, but had left by the time the writer arrived. Between the two world wars there was a great migration south from Scotland, and this brought the Carr family to Rookery Farm and the Steels to Manor Farm. In 1933 Harry Abel Berry came to Grange Farm, and the whole estate was owned and let by Lord Barrymore and later his daughter, the Honourable Mrs Dorothy Bell and her husband, Major Bertram (Billy).

The shape of the estate had not changed for nearly 1,500 years, it being possible to judge the age of the boundary hedges by the species they contain. In the original enclosure map in the County Record Office, the field boundaries are virtually identical to those on a drainage map of 1840, and were still relevant and identifiable 150 years after that. The record for 1245 shows the planting of the Grove Wood as a harbour for deer, and some of the spinneys at Grange Farm were put in as fox coverts in the mid-1800s. However, the basic outlines would still be recognised by someone who had known the place 1,000 years earlier. In fact, the old ditches and banks are still as they were before the Grove Wood was planted over them, as is the ridge and furrow field pattern in Dipslade Spinney.

The Bell family continued to visit Hamerton for two shooting visits a year, staying at The George at Buckden, and bringing a party of friends and relations who included at different times the prime minister of Northern Ireland (Mr Chichester Clark), Lord Dunraven, Lord St Germyns, and others not quite so smart. They were very well fed by Rita Berry at Church Farm for their shooting lunch (see her story later).

The eight farms and one smallholding carried three dairy herds in 1950, as well as six flocks of sheep and four herds of beef animals. Each farm had its own Horse Meadow, a relic from the very recent pre-war past, and everyone seemed to have a Barn Field and Home Close as well. When farms were amalgamated later on this caused some confusion with men and tractors going to work the wrong field, hence some rather unimaginative new names on the enlarged Grange Farm. There were potatoes, sugar beet, and even some Brussels sprouts grown, and agriculture employed at least twelve people, over and above the tenants and their families.

J

There was a full-time gamekeeper, Albert Spring (and before him Eric Goodwin's father, Freddie); the builder, Nobby Dodson, who kept a Jersey herd on the grass fields in the village, and we had our own road man, Freddy Cooper, who kept all the drainage grips within the parish boundaries clear, and an eye on everything else. His wife played the organ in church and a resident rector helped to complete the picture of a self-contained unit.

The Reverend Peter Weir had been a major in the King's African Rifles, and so brought a certain worldliness to the job; his son, also a soldier, was remembered with awe by children for his party trick of fire eating. Mrs Weir was a sweet person with very green fingers, and had created a lovely sunken garden with unusual irises. Under Canon Green, in the early part of the century, the rectory also boasted a fine pair of tennis courts, which lay between the copper beech tree and the church, but which have long since reverted to a rough children's play area.

When the Weirs retired, there followed the usual interregnum before the Sutton family was installed, and the Reverend Peter Sutton was to prove to be the last resident Rector of Hamerton. Although he, himself, had a background in the gas industry, he was fascinated by the old records, and became a local historian of great note and encyclopaedic knowledge. It is thanks to him that there is so much detail known about the neighbouring parishes and their inhabitants of years gone by, and to his meticulous indexing that this information is readily available in Huntingdon Record Office. Sadly, his wife Iris became seriously ill and died while the family were at Hamerton.

When the Bell family sold the estate in the late 1960s, it was decided that the old thatched cottages were going to handicap the sale, and some ten old houses were knocked down and used for farm roads. Among these was the home of Mrs Kath Joyce's mother-in-law near the village hall, and that of Mrs Elsie Radwell's parents, the Chesters, by the ford. Now even the ford has gone, in a major road-improvement scheme in the summer of 1968. The county surveyor congratulated himself and his team on a job well done with the comment "Well, you won't have any more flooding at Hamerton." The next twenty-four hours saw almost continuous rainfall, to total over four inches, and all that was to be seen of the major improvement was the handrail above the footpath.

To be fair to the owners and agent, the local authority had been gunning for them for many years: they were too low, too damp, too dark, etc., etc., and it was made clear to any would-be purchaser that the demolition order would be almost impossible to remove, and that the houses would never be passed for habitation again. Of course we now know different, from the few lovely old places that survive and look so good.

The lost houses have been replaced over the last twenty-five years and the population number remains much as it was, although the emphasis and dependence on agriculture has gone now, in common with that in every other rural community. Economics alone have dictated that farmers have had to expand, diversify, or retire from farming altogether, and the result is that Hamerton now supports a surprisingly wide range of other activities.

There is a garden-machinery business at Rookery Farm; a bridge-building contractor at Manor Farm, both run by the farming partners. There is a carpet warehouse and wholesaler and also a champagne importer at Church Farm; Dick Baxter's woodworking business has gone from strength to strength, now in a new purpose-built factory and run by his two sons — there are many wonderful pieces of their craftsmanship to be found around the district, such as the bell-tower arch in Catworth Church (and an even finer one at Stevenage). There are also some quality kitchens thanks to the Teals, who occupy what used to be the Baxter workshop, having rebuilt that and a house on the site.

Numerous members of the community now work from home (in financial services, editing, etcetera) linked to their offices by computer network. An equally large number seem to depart for far fields before dawn and reappear late, having fought legal battles, television producers, drama and other students, building developers, and many more; but we still have a lively, friendly community at the heart.

Charlie Slack 1942-2003

Charlie's family had lived in the Alconbury and Weston area for as long as anyone can remember, and for many years he and his brothers worked loyally and happily for Sylvia Rowley. I was very fortunate in that I needed an extra skilled man at exactly the same time as Mrs Rowley was cutting back, and that is how he and I came together. The third member of the team, Peter Hall, had joined in similar circumstances the year before that, and so from 1970 until I retired in 1996, the two were the mainstay of my business, and the reason that I felt no qualms at taking the odd job off the farm. In fact it gave food for thought when I came back to find that everything was exactly as I would have chosen, and often better! I was not indispensable after all. Much of what I have to say about Charlie applies to Peter also, as they were the epitome of the skilled farm worker: willing to tackle anything, expert at most aspects of stock or crop husbandry, and ready and quick to master new machinery or techniques however ludicrous their boss's idea might seem. They worked as if it was their own business on the line, made my interests theirs, and treated my young family as their own children. But it is Charlie who has died in tragic circumstances, and about him that I must now be more specific.

Apart from those aspects of his character, he had an unusually wide and varied range of interests and hobbies. He was an observant and knowledgeable naturalist, and spent happy hours watching, filming and video-recording wildlife in the woods and on the farm. His patience and attention to detail resulted in some outstanding programmes, ranging from the first courtship of a pair of kingfishers right through to the triumphant launch of the young on their first fishing trip. Another tale that was recorded ended in disaster when a grass snake devoured all the young whitethroats that he had been filming for weeks. He was so upset by this that he destroyed the film, being unable to bear watching its re-enactment. When Albert Spring retired as the gamekeeper, it seemed as if the job was tailored specially for Charlie. He was able to do his keeper's round before the breakfast break, and with one of the student helpers (usually his son Robert) covering for him on the tractor at feed times, he successfully reared and turned out over 1,000 birds. He put in countless hours of his own free time as well. When the shooting season came along, after the arable rush was over, he entered into the spirit wholeheartedly, and built up a great network of beaters and helpers. The result was sport that was both testing and fun, and provided a huge amount of enjoyment to the participants, down to the smallest and most unsuitable terrier dog!

Another aspect of his love of bird life was to collect the corpses of those unfortunate enough to fly into glass windows or traffic, and have them mounted professionally at considerable expense. The idea may not be to everyone's taste, but they are works of art, and he took pride in his rare collection, which has some fabulous specimens in it. His fish, some as big as sharks, are also quite a sight, as they fill the small pond in the garden, and threaten to swim off down to the Great Ouse and freedom whenever the Alconbury Brook floods.

His knowledge of the planes that flew from the base at Alconbury was encyclopaedic, and he could identify anything flying by its tweet or engine note. When the A10 Tank Busters were practising before the last Gulf War, they used our tractors and combines as targets, and I was never sure that the pilots had spotted the second one, as they practically took the roof off the combine. Charlie really loved this, although less brave mortals like myself, and Peter too I suspect, worried slightly. With his friendships on the base, at the Rod and Gun Club, and his own observation, he seemed to have access to some highly classified information: he knew where the U2 spy planes were going and what they were looking at before the American Air Force did themselves.

We have some happy snapshot memories of him in other circumstances; impervious to electric shocks on the fencer, he time and again persuaded one of us that we had switched off the current, only to get a hell of a kick. Finally we got him to agree that there was a faint tickle in the system by taking his boots off and standing him in a puddle.

He would save his best shave (sometimes his only one of the week) for Friday night and the special Chinese meal. He was 'wet-headed' — however bright and set fair, he had heard an ominous forecast of worse to come — I blame Fred and Radio Norfolk for that.

He did have an endearing quality of enjoying the role of bearer of bad news: he would come up with a huge grin and say, "You'll never guess what's happened!" or "You're not going to like this, boss!": it was obviously a moment to be savoured, and I racked my brains to work out the most probable disaster. I knew it would be something like a grain trailer tipped into a ditch, or the harvest student removing the barn roof by driving out with the trailer elevated.

He had recently started taking Lena abroad on holiday, but for years she was lucky to get further than Minsmere, Snettisham and other bird reserves.

As sprayers became ever more complex and the chemicals more unpronounceable, we all struggled, and Charlie produced some amazing versions of Ciba-Geigy's latest wonder. He was always spot on with his dose rates, and mixes, whatever our agronomist, David Boothroyd, threw at us. I felt he really loved his sprayer, and he had apparently turned down the offer of a new one, as being disloyal to his old friend. In a way his end, however untimely, was perhaps the scenario he would have written for himself. He was in the open fields, doing the job he did best, in the machine that only he had driven from new. But what a terrible shock for his family and friends, for Peter who found him dead at the wheel with the engine still running, and worst of all for his son, Robert, and Lena, who had the news broken to them when they returned from work! Our deepest sympathy goes out to them and all his family.

Michael Halford

Steeple Gidding News

Geoffrey Butcher, died 14th March 2003
When I and my family first moved into Grange Farm, in October 1959, we did not have any local contacts, and were amazed at the number of our new neighbours who went out of their way to make us welcome. Among the first was Mrs Butcher, at the Old Rectory, with her son Geoffrey, where I now live. They were kindness and hospitality personified, and Emily Butcher was a marvellous and stylish hostess. Of course I never called her by her Christian name; she was a bit too daunting for that, but that didn't stop her from showing great kindness to many people, and our first visit here was to a magnificent Sunday dinner. She managed to keep this church open for a monthly service long after she was the only member of the congregation, so in awe of her were the bishop, the archdeacon, the rural dean and the vicar of The Giddings, and when she passed away at the grand age of nintey-three no time was lost in closing the church.

Geoff was a much quieter character by comparison with his mother; he was always calm and steady, unfailingly polite and encouraging to others struggling with a farming problem or their shooting, or anything else that life threw at them. I never heard him say an unkind word about anyone; the positive side would always come first, and he had a lovely understated sense of humour bubbling within him. Generous, kind and patient with others, he in fact had some periods of great pain and physical suffering with stomach cancer over thirty years ago, which he endured without a moan. He must have been on his second or third lease of life after he had been through that lot..

He and his brother Boyd had farmed together, Manor Lodge and Coldharbour Farms, Steeple Gidding for fifteen years until they retired in 1968. In fact Geoffrey and Mrs Butcher lived at Manor Lodge for a while until Hill Royde came up for sale, when they bought it and reclaimed its original name. Boyd and his wife returned to South

Africa, and Geoffrey moved down to Dorset where he lived until increasing infirmity brought him back to be looked after lovingly by his niece, Janey. He really enjoyed being taken round in her very smart open Jaguar, and visited his old friends as long as he was able. Always a countryman from his boots up, he was a fine shot, selective in ignoring anything low or that could conceivably be flying towards his neighbouring gun, and giving the credit for anything shared to the other. A regular guest at Hamerton and Conington, as well as other local shoots and his old haunts in Derbyshire; he was a good friend of the late Albert Spring, and also of John Heathcote, and will be remembered with affection by many more, not least yours truly. Robbie Steel, his godson, regularly called on him in Dorset, and often took his son Ben over to see him at Hartford.

He had become profoundly deaf in his last year, and was in hospital following a bad fall which broke his leg in two places. A final irony was that at the time of his death in Hinchingbrooke, I should have been on the golf course with Janey, John Simpson (another of Geoff's old friends), and his retired doctor, Bob Berrington. Our sincere sympathy goes out to Janey, who will miss him more than any.

Church Repairs

For the past month the church has been shrouded and sheeted with corrugated iron, while the roof has been re-leaded, the stonework repaired, and the interior painted. There is also a plan to put an access ladder to the bell chamber, but this is then going to present a hazard to the visiting public, and it is said that the tower must be kept locked! I am still arguing the case for access to the bell ropes, as many children (and grown-ups) have enjoyed playing tunes on the three bells, and it is good to hear them ringing out and bringing life to the old place.

The work should be completed by the end of this month, or soon after, and the church ready in time for the spring rush of visitors!

Michael Halford

HAMERTON TENNIS TOURNAMENT 2002

Once again this event, held on Howard and Fiona Anderson's court, provided a huge amount of fun and noise, and even raised £70 for the church funds. Only five pairs (or to be precise, four pairs and a trio) took part this year; for various reasons some of our regulars were unable to play, but nevertheless the light was fading fast by the time the winners had finished with their unseemly crowing and gloating.

Those who have watched and wondered at the level of crowd participation during the Davis Cup matches would not have been quite so overcome as the players themselves at Hamerton. The cries of encouragement and otherwise were almost continuous; the unofficial umpires and scorers were frequently at variance with the facts and with each other, but made up for that by sheer volume, and small spectators in the De Grussa's garden next door helped by asking, "What's your name? What are you doing?" whilst balanced up a ladder. The net result of all this was to reduce the potential ability of the star player to something nearer the average, which was bad luck on her, but better for the rest of us!

Amelia, who partnered Sarah Anderson, is quite clearly a very good player indeed, and had captained the school team at Kimbolton this summer. Most of the time she tried to pull her punches to avoid unnecessary cruelty to the opposition, and as a result made more than her ration of unforced errors. When she did murder the ball overhead, though, it stayed murdered, and some of her smashes nearly cleared the giant plane tree outside the church. The effect of this was not the only reason that of the five boxes of balls that started the day with four in each, only three balls we to be seen at the end of the afternoon.

The pair had no difficulty in reaching the final, but there met the local secret weapon, the combination of J. V. and H. Swales; Vernon and Hannah had started the afternoon in quite a low key, behaving with exemplary decorum to their opponents and each other. Had the leopard changed his spots, we wondered? The answer was forthcoming as the heat of battle intensified: no, he had not changed his spots. One has to admire a man who is no longer exactly in the first flush of youth (not wishing to cause offence in saying that), who is prepared to shout "MINE" in a stentorian bellow, when the ball is way beyond the far tramlines, and what's more rush over and return it time and time again. (This attitude was in sharp contrast to a recent tournament which I helped run for Macmillan Cancer Relief, when most of the men specifically demanded a young, nubile and athletic partner to do all the hard work.) Magnificent! We all cheered, although Hannah herself seemed a trifle miffed at the constant poaching of her side of the court. In spite of that the tactics worked, and the father/ daughter pair beat Sarah and Amelia 3-2 to win the bigger boxes of chocolates.

The also-rans, included another, quieter, Swales pair of Jane and Emily, who acquitted themselves with dignity and some success. The trio of Howard, Brian and Rachael, rotated on and off the substitutes bench and got themselves and everyone else confused, but produced some amazing tennis (using the term in its broadest sense). Sometimes the rallies were lightning fast and we could only gulp at the sheer speed of Brian's reflexes. At other times the connection of racquet and ball was absent altogether, but you never knew which was coming. (Brian put it all down to badminton!) Howard's 8th-Army shorts were once again a feature, and as for the style of his play, poetry in motion best describes it: there were pirouettes, triple axels and other amazing moves which made Nureyev's efforts of the past pale. Had Howard trained at the Royal School of Ballet, we wondered, and why wasn't he wearing a tutu? The tennis ball remained unimpressed by all this in the meantime and moved on.

Your correspondent, being partnered by the hostess of the event, was resigned to not being allowed to win, and so was not unduly disappointed. He did nevertheless enjoy the lemonade, tea and biscuits, washed down with generous helpings of wine, and the thought occurs that this possibly accounted for some of the variability in the actual tennis. In short, an excellent time was had by all, and three cheers for everyone — particularly Howard and Fiona! *MJH*

PS Later reports claim that all bar three of the balls returned to base finally.
PPS Brian Dutton could barely walk when next seen in public.

Hamerton Activity

Cars parked all down the street; a gentle smoke cloud drifting over the village, along with mouth-watering smell of roast something; the happy cries of children banging each other's heads as they bound around the bouncy castle. It can only mean one thing — the Summer Barbeque is here again! And How!

To get in through the gate one had to run the gauntlet of Rita, Sonya and Mary. To compare them to the three witches from *Macbeth* would be a little unfair, but there was a faint echo of *Arsenic and Old Lace* as they smiled sweetly at each new victim, and relieved them of their money. Perhaps 'Come into my parlour, said the spider to the fly' would be a better simile. Anyway, they collected a huge stack of gold at the gate, and ensured that the evening was going to be a financial success whatever went on inside. There were also some mouthwatering cakes, and the strongest selling point was that the shortbread had been handcrafted by Alan Steel. Delicious, too! Is there no limit to the man's talent?

Numbers present were hard to pin down, owing to the fluid state of the battle, but as over £800 was collected on the gate, at £6 per head, there must have been in the region of 130 people there. Would there be enough to eat/drink? Of course there was, and brilliantly cooked as usual.

Brian, John and Kevin took it in turns to baste and roast themselves on the grill, and Brian in particular seemed to have spent most of the day trying to keep the temperamental gas alight. By the time the paying customers arrived he looked like a cross between the Naked Chef and Spartacus, whereas John and Kevin were more I'm A Celebrity . . .Get Me Out Of Here! from the Australian bush and wondering who was going to have to tackle the killer ants next.

Who else did what? Well the tent obviously puts itself up by now, but there may have been a Swales or two, Spartacus and Crocodile Dundee involved in some small degree. Crocodile Dundee, with his favoured nation status at The Green Man, organized the beer, and the bar seemed to be staffed by all the serious drinkers of this parish, but the Anderson family were loath to be parted from it for long. Howard's role was to rush around looking frightfully busy, and WORRIED. A hard act to follow!

The team of good fairies once again came up with salads and side dishes, raffle prizes and sold tickets, and it is more a case of who didn't do anything than who did. (Well, your correspondent didn't for one, but he had to sharpen his pencil.)

The producer/director and general enabler of the whole fair was Maxine Blades, God bless her. There was the added bonus to her of Kevin's birthday next day, needing no more input than to summon the whole world to help finish up the dregs; as the dregs lasted from 11 a.m. to 7 p.m. I think it can be taken that there was a slight element of over-ordering.

So once again, a resoundingly happy and successful event, with nearly £800 profit for the village hall and a great time had by all. THANK YOU and WELL DONE!

Steeple Gidding Graves and Their Stories

The oldest coffin lid is standing up in the South-East corner of the church, but we know little about it or the person whose remains were in it; it is said to date from the thirteenth century, but that is it! As Steeple Gidding was ceded to Ramsey Abbey before the 1066 conquest, it is probably a senior member of that foundation who was honoured by the elaborate stone coffin.

After that there is a long gap in stone memorials; suitable material was hard to come by and expensive. Skilled stone carvers were rare and, when found, proved to be virtually illiterate. From the way that they split names over two lines of script (see Elizabeth Sherrard's memorial under the window of the south aisle) and made spelling mistakes, they clearly did not understand what they were carving.

The Sherrard tablet: *Hear lieth y body of Elzeb*
eth Shard the dater of Mich
ael Shard was buried June 30th
1682

Her husband was buried in 1689, but there is no stone to him, although he was churchwarden. The year after he died his daughter married Jeffrey Hull, and that is the last record of the family in the village.

Another old, beautifully carved and preserved tombstone stands outside the east end of the church.

It reads: *Here lieth the body of William Ruff And was beried March the eigt day 1684 Ano Dom.*

All we know about him is that he is recorded as a yeoman but this is the only mention of the Ruffs of Steeple Gidding. It is a local name, though, and there is a Ruff charity fund for the poor of Winwick, and other mentions of the family round about.

The earliest decipherable memorials are in the church, fairly understandably.

The Cottons

This lies at the east end of the south aisle: *Hereunder this stone resteth the body of Thomas Cotton, owner of this manor of Steeple Gidding, second son of Thomas Cotton of Conington Esq. Hee died the first day of April Ano Dni 1640*

The Cotton family owned substantial estates in this area, and were major powers in the land in Tudor times. Apart from the 1,066 acres of Steeple Gidding, with a fine residence and formal garden, they had their main seat at Conington Castle, just across the A1. Another branch of the family owned the other Conington estate, which is only just over the original Hunts/Cambs border, and Madingley Hall, near Cambridge.

Sir Robert is credited with making the bequest, which later formed the foundation of the British Museum, of his wonderful library and collection of ancient documents. His wife's family also were noted antiquarians, and their collection, the Annesley papers, was gifted to the nation at the same time. The family claimed descent from Robert the Bruce, King of Scotland, as witness this almost excessively detailed account of the connections. 'Methinks she protesteth too much', comes to mind, but may be an unfair judgement. There is after all Bruce's Castle Farm on the Conington Estate, and for generations the Bruces (also spelt Brous/Brus) were also Earls of Huntingdon.

The inscription on the large tablet in the north wall reads as follows: *Near this place lyeth the body of Sir John Cotton, Bart, who departed this life March the 27th 1752 in the 65th year of his age, son of Sir Robert Cotton of this parish, Bart, who died the 12th day of July 1749 aged 81, and who is here buried in the same vault in which are likewise the remains of his son and grandchildren.*

He was Lord of the Manor of Steeple Gidding, Saltree, Denton and Conington in this county, and of Stretton in the county of Bedford, which two last mentioned, with those of Glatton and Holme he became possessed of by his cousin Sir John Cotton, Bart, who died at London February 5th 1731, and was privately interred in the burying ground of Saint George the Martyr according to the express directions of his will.

He married Jane, daughter of Sir Robert Burdett of Formack in

Derbyshire and of Bramcote in Warwick, Bart, by whom he had two sons and seven daughters.

His eldest son Robert, born February 16th 1711/12 was a child of uncommon hopes and of such a disposition as seemed to promise in him a succession worthy of his father, but it pleased God to take him February 1715/16 when he was not full four years old. His younger son John whose education was chiefly owing to the care of his father and who has made ample amends for that care by his great improvement and accomplishment died November 15th 1739 to the irreparable loss of his family in the 25th year of his age, having lived long enough to be universally regretted.

Of his daughters three, viz Mary, Mary and Hester, died young, and Jane, Elizabeth-Stuart, Frances and Mary survive.

In Sir John ended the male line of the ancient, honourable and loyal family of Bruce-Cottons, so called from Robert de Brus, Great Grandfather to King Robert de Brus, ancestor to the Royal family of Stuart which Robert the Elder married Isabella, grand daughter to David the King of Scotland. By this marriage his successors inherited the Lordship of the Manor of Conington and the Royal arms of Scotland, both which descended by an heiress of Brus to their families of Wessenham and from thence after the like manner of descent passed into this family. The dignity of Baronet was conferred on Sir Robert Cotton of Conington, founder of the Cotton library, by King James 1st June 9th 1611. The father, son and grandson constantly adhered both in principle and practice to their duty and the true interest of their country and were in no way inferior to their ancestors all whose virtues they inherited.

The Lord grant unto them that they may find mercy in the Lord in that day; (2 Tim 1.18)

This monument was erected by Dame Jane Cotton. MDCCLII

It would seem likely that all the Cotton children are buried in the vault in the centre of the church, but there are two gravestones outside the chancel, in a quiet corner by themselves, which are difficult to read, but which could be the two infant Marys who died. The Mary who survived is commemorated by an elaborate plaque in the chancel, as the wife of Roger Kinyon (Kenyon). A fulsome eulogy has proved difficult to decipher in its entirety.

Wil And Elianor Hanger!

There is a beautifully carved and preserved old gravestone near the front gate: *"Here lieth ye body of Elianor, wife of William Hanger*

who departed this life December y 5th 1718 in y 74 yeare of her age"
There was no sign of a stone to Wil until some tidying up in the churchyard revealed a heap of stone, eminently suitable for making a rockery, in a jumbled heap in the hedge bottom. Only when the rockery was under construction did the letters W H reveal themselves, and it did not require more than a quick look in the parish records to establish the identity of the owner.

The story that unfolded from the records is intriguing, as Widow Bright (Elianor) is found to have married Wil Hanger only one month after her first husband had died. (Richard Bright's death is recorded on 22nd November 1688.)

The Brite/Bright family had been around for a long time, and are noted on the wonderful Hausted map of 1647 as smallholders at the end of the Steeple Gidding lane — near where the telephone box now stands. Further back in 1603, Arthur Brite married Ursula Loufe only four months before their son, Edward, was born.

By 1632 Edward was marrying Anne Pasheler, just beating the arrival of their first-born, Ursula, by six weeks. Their second child, our Richard, was born the next year and baptised on 16th February.

In 1666 (the year of the Great Fire of London) Richard married Ellin Smith of Hamerton, and over the next twelve years they reared four daughters and two sons. Unusually for those days, all survived infancy until their youngest, Mary, died at the age of seven in 1685.

The family had established a history of living to a good age. Anne (nee Pasheler) only died in 1669, and her husband Edward Bright in 1678, so when Richard himself died ten years after his father it was against previous form.

All the families mentioned in this little saga crop up regularly in other contexts. For instance, Richard Bright's name appears in a court case when he had some cattle stolen; Simon Smith, father of Ellen (Elianor) was a witness to the transfer of glebe land at the time of the enclosure of Hamerton. The Hangers were mainly from Abbots Ripton and Stukeley, but Martha, infant daughter of William Hanger 2nd (or 3rd), was christened and buried at Steeple Gidding in 1705.

Did the Bright children disapprove of their mother's very quick re-marriage, and quietly dispose of their stepfather's simple gravestone into the hedge? They certainly made sure that their memorial stone was worthy of her. There is a regular pattern of what would today be considered unseemly haste in re-marrying after the death of a husband or wife, but times were hard and for a widow to bring up a family without a provider was no worse than the predicament of a widowed husband with young children.